The DOORS

of PERCEPTION

&

HEAVEN *and*

HELL

BOOKS BY ALDOUS HUXLEY

Novels

The Genius and the Goddess

Ape and Essence

Time Must Have a Stop

After Many a Summer Dies
the Swan

Eyeless in Gaza

Point Counter Point

Those Barren Leaves

Antic Hay

Crome Yellow

Brave New World

Island

Essays and Belles Lettres

Brave New World Revisited

Tomorrow and Tomorrow and
Tomorrow

Heaven and Hell

The Doors of Perception

The Devils of Loudun

Themes and Variations

Ends and Means

Texts and Pretexts

The Olive Tree

Music at Night

Vulgarity in Literature

Do What You Will

Proper Studies

Jesting Pilate

Along the Road

On the Margin

Essays New and Old

The Art of Seeing

The Perennial Philosophy

Science, Liberty and Peace

Short Stories

Collected Short Stories

Brief Candles

Two or Three Graces

Limbo

Little Mexican

Mortal Coils

Biography

Grey Eminence

Poetry

The Cicadas

Leda

Travel

Beyond the Mexique Bay

Drama

Mortal Coils—A Play

The World of Light

The Discovery, Adapted from
Francis Sheridan

Selected Works

Rotunda

The World of Aldous Huxley

The Doors
of Perception

&

Heaven and
Hell

ALDOUS
HUXLEY

HARPER**PERENNIAL** ● MODERN**CLASSICS**

NEW YORK ● LONDON ● TORONTO ● SYDNEY ● NEW DELHI ● AUCKLAND

A hardcover edition of *The Doors of Perception* was published in 1954 by Harper & Brothers, Publishers. A hardcover edition of *Heaven and Hell* was published in 1956 by Harper & Brothers, Publishers.

P.S.™ is a trademark of HarperCollins Publishers.

HarperCollins books may be purchased for educational, business, or sales promotional use. For information, please e-mail the Special Markets Department at SPsales@harpercollins.com.

First Perennial Library edition published 1990.
First Perennial Classics edition published 2004.
First Harper Perennial Modern Classics edition published 2009.

Library of Congress Cataloging-in-Publication Data is available upon request.

ISBN 978-0-06-172907-2

23 24 25 26 27 LBC 36 35 34 33 32

For M

"If the doors of perception were cleansed every thing would appear to man as it is, infinite."

—WILLIAM BLAKE

The DOORS
of PERCEPTION

It was in 1886 that the German pharmacologist, Louis Lewin, published the first systematic study of the cactus, to which his own name was subsequently given. *Anhalonium Lewinii* was new to science. To primitive religion and the Indians of Mexico and the American Southwest it was a friend of immemorially long standing. Indeed, it was much more than a friend. In the words of one of the early Spanish visitors to the New World, "they eat a root which they call peyote, and which they venerate as though it were a deity."

Why they should have venerated it as a deity became apparent when such eminent psychologists as Jaensch, Havelock Ellis and Weir Mitchell began their experiments with mescalin, the active principle of peyote. True, they stopped short at a point well this side of idolatry; but all concurred in assigning to mescalin a position among drugs of unique distinction. Administered in suitable doses, it changes the quality of consciousness more profoundly and yet is less toxic than any other substance in the pharmacologist's repertory.

Mescalin research has been going on sporadically ever since the days of Lewin and Havelock Ellis. Chemists have not merely isolated the alkaloid; they have learned how to synthesize it, so that the supply no longer depends on the sparse and intermittent crop of a desert cactus. Alienists have dosed themselves with mescalin in the hope thereby of coming to a better, a first-hand, understanding of their patients' mental processes. Working unfortunately upon too few subjects within too narrow a range of circumstances, psychologists have observed and catalogued some of the drug's more striking effects. Neurologists and physiologists have found out something about the mechanism of its action upon the central nervous system. And at least one professional philosopher has taken mescalin for the light it may throw on such ancient, unsolved riddles as the place of mind in nature and the relationship between brain and consciousness.

There matters rested until, two or three years ago, a new and perhaps highly significant fact was observed.*

* See the following papers: "Schizophrenia. A New Approach." By Humphry Osmond and John Smythies. *Journal of Mental Science.* Vol. XCVIII. April, 1952.

"On Being Mad." By Humphry Osmond. *Saskatchewan Psychiatric Services Journal.* Vol. I. No. 2. September, 1952.

Actually the fact had been staring everyone in the face for several decades; but nobody, as it happened, had noticed it until a young English psychiatrist, at present working in Canada, was struck by the close similarity, in chemical composition, between mescalin and adrenalin. Further research revealed that lysergic acid, an extremely potent hallucinogen derived from ergot, has a structural biochemical relationship to the others. Then came the discovery that adrenochrome, which is a product of the decomposition of adrenalin, can produce many of the symptoms observed in mescalin intoxication. But adrenochrome probably occurs spontaneously in the human body. In other words, each one of us may be capable of manufacturing a chemical, minute doses of which are known to cause profound changes in consciousness. Certain of these changes are similar to those which occur in that most characteristic plague of the twentieth century, schizophrenia. Is the mental disorder due to a chemical disorder? And is the chemical disorder due, in its turn, to psychological distresses affecting the

"The Mescalin Phenomena." By John Smythies. *The British Journal of the Philosophy of Science.* Vol. III. February, 1953.

"Schizophrenia: A New Approach." By Abram Hoffer, Humphry Osmond and John Smythies. *Journal of Mental Science.* Vol. C. No. 418. January, 1954.

Numerous other papers on the biochemistry, pharmacology, psychology and neurophysiology of schizophrenia and the mescalin phenomena are in preparation.

11

adrenals? It would be rash and premature to affirm it. The most we can say is that some kind of a *prima facie* case has been made out. Meanwhile the clue is being systematically followed, the sleuths—biochemists, psychiatrists, psychologists—are on the trail.

By a series of, for me, extremely fortunate circumstances I found myself, in the spring of 1953, squarely athwart that trail. One of the sleuths had come on business to California. In spite of seventy years of mescalin research, the psychological material at his disposal was still absurdly inadequate, and he was anxious to add to it. I was on the spot and willing, indeed eager, to be a guinea pig. Thus it came about that, one bright May morning, I swallowed four-tenths of a gram of mescalin dissolved in half a glass of water and sat down to wait for the results.

We live together, we act on, and react to, one another; but always and in all circumstances we are by ourselves. The martyrs go hand in hand into the arena; they are crucified alone. Embraced, the lovers desperately try to fuse their insulated ecstasies into a single self-transcendence; in vain. By its very nature every embodied spirit is doomed to suffer and enjoy in solitude. Sensations, feelings, insights, fancies—all these are private and, except through symbols and at second hand, incommunicable. We can pool information about experiences, but never the experiences themselves. From family to nation, every human group is a society of island universes.

Most island universes are sufficiently like one another to permit of inferential understanding or even of mutual empathy or "feeling into." Thus, remembering our own bereavements and humiliations, we can condole with others in analogous circumstances, can put ourselves (always, of course, in a slightly Pickwickian sense) in their places. But in certain cases communication between universes is incomplete or even nonexistent. The mind is its own place, and the places inhabited by the insane and the exceptionally gifted are so different from the places where ordinary men and women live, that there is little or no common ground of memory to serve as a basis for understanding or fellow feeling. Words are uttered, but fail to enlighten. The things and events to which the symbols refer belong to mutually exclusive realms of experience.

To see ourselves as others see us is a most salutary gift. Hardly less important is the capacity to see others as they see themselves. But what if these others belong to a different species and inhabit a radically alien universe? For example, how can the sane get to know what it actually feels like to be mad? Or, short of being born again as a visionary, a medium, or a musical genius, how can we ever visit the worlds which, to Blake, to Swedenborg, to Johann Sebastian Bach, were home? And how can a man at the extreme limits of ectomorphy and cerebrotonia ever put himself in the place of one at the limits of endomorphy and viscerotonia, or, except within certain circumscribed areas, share the feelings of one who stands at the limits of mesomorphy and somatotonia? To the unmitigated behaviorist such questions, I suppose, are meaningless. But for those who theoretically believe what in practice they know to be true—namely, that there is an inside to experience as well as an outside—the problems posed are real problems, all the more grave for being, some completely insoluble, some soluble only in exceptional circumstances and by methods not available to everyone. Thus, it seems virtually certain that I shall never know what it feels like to be Sir John Falstaff or Joe Louis. On the other hand, it had always seemed to me possible that, through hypnosis, for example, or autohypnosis, by means of systematic meditation, or else by taking the appropriate drug, I might so change my ordinary mode of consciousness as to be able to know, from the inside, what the visionary, the medium, even the mystic were talking about.

From what I had read of the mescalin experience I was convinced in advance that the drug would admit me, at least for a few hours, into the kind of inner world described by Blake and Æ. But what I had expected did not happen. I had expected to lie with my eyes shut, looking at visions of many-colored geometries, of animated architectures, rich with gems and fabulously lovely, of landscapes with heroic figures, of symbolic dramas trembling perpetually on the verge of the ultimate revelation. But I had not reckoned, it was evident, with the idiosyncrasies of my mental make-up, the facts of my temperament, training and habits.

I am and, for as long as I can remember, I have always been a poor visualizer. Words, even the pregnant words of poets, do not evoke pictures in my mind. No hypnagogic visions greet me on the verge of sleep. When I recall something, the memory does not present itself to me as a vividly seen event or object. By an effort of the will, I can evoke a not very vivid image of what happened yesterday afternoon, of how the Lungarno used to look before the bridges were destroyed, of the Bayswater Road when the only buses were green and tiny and drawn by aged horses at three and a half miles an hour. But such images have little substance and absolutely no autonomous life of their own. They stand to real, perceived objects in the same relation as Homer's ghosts stood to the men of flesh and blood, who came to visit them in the shades. Only when I have a high temperature do my mental images come to independent life. To those in whom the faculty of visualization is strong my inner world must seem curiously drab, limited and uninteresting. This was the world—a poor thing but my own—which I expected to see transformed into something completely unlike itself.

The change which actually took place in that world was in no sense revolutionary. Half an hour after swallowing the drug I became aware of a slow dance of golden lights. A little later there were sumptuous red surfaces swelling and expanding from bright nodes of energy that vibrated with a continuously changing, patterned life. At another time the closing of my eyes revealed a complex of gray structures, within which pale bluish spheres kept emerging into intense solidity and, having emerged, would slide noiselessly upwards, out of sight. But at no time were there faces or forms of men or animals. I saw no landscapes, no enormous spaces, no magical growth and metamorphosis of buildings, nothing remotely like a drama or a parable. The other world to which mescalin admitted me was not the world of visions; it existed out there, in what I could see with my eyes open. The great change was in the realm of objective fact. What had happened to my subjective universe was relatively unimportant.

I took my pill at eleven. An hour and a half later, I was sitting in my study, looking intently at a small glass vase. The vase contained only three flowers—a full-blown Belle of Portugal rose, shell pink with a hint at every petal's base of a hotter, flamier hue; a large magenta and cream-colored carnation; and, pale purple at the end of its broken stalk, the bold heraldic blossom of an iris. Fortuitous and provisional, the little nosegay broke all the rules of traditional good taste. At breakfast that morning I had been struck by the lively dissonance of its colors. But that was no longer the point. I was not looking now at an unusual flower arrangement. I was seeing what Adam had seen on the morning of his creation—the miracle, moment by moment, of naked existence.

"Is it agreeable?" somebody asked. (During this part of the experiment, all conversations were recorded on a dictating machine, and it has been possible for me to refresh my memory of what was said.)

"Neither agreeable nor disagreeable," I answered. "It just *is*."

Istigkeit—wasn't that the word Meister Eckhart liked to use? "Is-ness." The Being of Platonic philosophy—except that Plato seems to have made the enormous, the grotesque mistake of separating Being from becoming and identifying it with the mathematical abstraction of the Idea. He could never, poor fellow, have seen a bunch of flowers shining with their own inner light and all but quivering under the pressure of the significance with which they were charged; could never have perceived that what rose and iris and carnation so intensely signified was nothing more, and nothing less, than what they were—a transience that was yet eternal life, a perpetual perishing that was at the same time pure Being, a bundle of minute, unique particulars in which, by some unspeakable and yet self-evident paradox, was to be seen the divine source of all existence.

I continued to look at the flowers, and in their living light I seemed to detect the qualitative equivalent of breathing—but of a breathing without returns to a starting point, with no recurrent ebbs but only a repeated flow from beauty to heightened beauty, from deeper to ever deeper meaning. Words like "grace" and "transfiguration" came to my mind, and this, of course, was what, among other things, they stood for. My eyes traveled from the rose to the carnation, and from that feathery incandescence to the smooth scrolls of sentient amethyst which were the iris. The Beatific Vision, *Sat Chit Ananda*, Being-Awareness-Bliss—for the first time I understood, not on the verbal level, not by inchoate hints or at a distance, but precisely and completely what those prodigious syllables referred to. And then I remembered a passage I had read in one of Suzuki's essays. "What is the Dharma-Body of the Buddha?" ("The Dharma-Body of the Buddha" is another way of saying Mind, Suchness, the Void, the Godhead.) The question is asked in a Zen monastery by an earnest and bewildered novice. And with the prompt irrelevance of one of the Marx Brothers, the Master answers, "The hedge at the bottom of the garden." "And the man who realizes this truth," the novice dubiously inquires, "what, may I ask, is he?" Groucho gives him a whack over the shoulders with his staff and answers, "A golden-haired lion."

It had been, when I read it, only a vaguely pregnant piece of nonsense. Now it was all as clear as day, as evident as Euclid. Of course the Dharma-Body of the Buddha was the hedge at the bottom of the garden. At the same time, and no less obviously, it was these flowers, it was anything that I—or rather the blessed Not-I, released for a moment from my throttling embrace—cared to look at. The books, for example, with which my study walls were lined. Like the flowers, they glowed, when I looked at them, with brighter colors, a profounder significance. Red books, like rubies; emerald books; books bound in white jade; books of agate; of aquamarine, of yellow topaz; lapis lazuli books whose color was so intense, so intrinsically meaningful, that they seemed to be on the point of leaving the shelves to thrust themselves more insistently on my attention.

"What about spatial relationships?" the investigator inquired, as I was looking at the books.

It was difficult to answer. True, the perspective looked rather odd, and the walls of the room no longer seemed to meet in right angles. But these were not the really important facts. The really important facts were that spatial relationships had ceased to matter very much and that my mind was perceiving the world in terms of other than spatial categories. At ordinary times the eye concerns itself with such problems as *Where?—How far? —How situated in relation to what?* In the mescalin experience the implied questions to which the eye responds are of another order. Place and distance cease to be of much interest. The mind does its perceiving in terms of intensity of existence, profundity of significance, relationships within a pattern. I saw the books, but was not at all concerned with their positions in space. What I noticed, what impressed itself upon my mind was the fact that all of them glowed with living light and that in some the glory was more manifest than in others. In this context position and the three dimensions were beside the point. Not, of course, that the category of space had been abolished. When I got up and walked about, I could do so quite normally, without misjudging the whereabouts of objects. Space was still there; but it had lost its predominance. The mind was primarily concerned, not with measures and locations, but with being and meaning.

And along with indifference to space there went an even more complete indifference to time.

"There seems to be plenty of it," was all I would answer, when the investigator asked me to say what I felt about time.

Plenty of it, but exactly how much was entirely irrelevant. I could, of course, have looked at my watch; but my watch, I knew, was in another universe. My actual experience had been, was still, of an indefinite duration or alternatively of a perpetual present made up of one continually changing apocalypse.

From the books the investigator directed my attention to the furniture. A small typing table stood in the center of the room; beyond it, from my point of view, was a wicker chair and beyond that a desk. The three pieces formed an intricate pattern of horizontals, uprights and diagonals—a pattern all the more interesting for not being interpreted in terms of spatial relationships. Table, chair and desk came together in a composition that was like something by Braque or Juan Gris, a still life recognizably related to the objective world, but rendered without depth, without any attempt at photographic realism. I was looking at my furniture, not as the utilitarian who has to sit on chairs, to write at desks and tables, and not as the cameraman or scientific recorder, but as the pure aesthete whose concern is only with forms and their re-

lationships within the field of vision or the picture space.
But as I looked, this purely aesthetic, Cubist's-eye view
gave place to what I can only describe as the sacramental
vision of reality. I was back where I had been when I
was looking at the flowers—back in a world where everything shone with the Inner Light, and was infinite in its
significance. The legs, for example, of that chair—how
miraculous their tubularity, how supernatural their polished smoothness! I spent several minutes—or was it
several centuries?—not merely gazing at those bamboo
legs, but actually *being* them—or rather being myself in
them; or, to be still more accurate (for "I" was not involved in the case, nor in a certain sense were "they")
being my Not-self in the Not-self which was the chair.

Reflecting on my experience, I find myself agreeing
with the eminent Cambridge philosopher, Dr. C. D.
Broad, "that we should do well to consider much more
seriously than we have hitherto been inclined to do the
type of theory which Bergson put forward in connection
with memory and sense perception. The suggestion is
that the function of the brain and nervous system and
sense organs is in the main *eliminative* and not productive. Each person is at each moment capable of remembering all that has ever happened to him and of perceiving
everything that is happening everywhere in the universe.
The function of the brain and nervous system is to pro-

tect us from being overwhelmed and confused by this mass of largely useless and irrelevant knowledge, by shutting out most of what we should otherwise perceive or remember at any moment, and leaving only that very small and special selection which is likely to be practically useful." According to such a theory, each one of us is potentially Mind at Large. But in so far as we are animals, our business is at all costs to survive. To make biological survival possible, Mind at Large has to be funneled through the reducing valve of the brain and nervous system. What comes out at the other end is a measly trickle of the kind of consciousness which will help us to stay alive on the surface of this particular planet. To formulate and express the contents of this reduced awareness, man has invented and endlessly elaborated those symbol-systems and implicit philosophies which we call languages. Every individual is at once the beneficiary and the victim of the linguistic tradition into which he has been born—the beneficiary inasmuch as language gives access to the accumulated records of other people's experience, the victim in so far as it confirms him in the belief that reduced awareness is the only awareness and as it bedevils his sense of reality, so that he is all too apt to take his concepts for data, his words for actual things. That which, in the language of religion, is called "this world" is the universe of reduced awareness, expressed, and, as it were, petrified by language. The various "other worlds," with which human beings erratically make contact are so many elements in the totality of the awareness belonging to Mind at Large. Most people, most of the time, know only what comes through the reducing valve and is consecrated as genuinely real by the local language. Certain persons, however, seem to be born with a kind of by-pass that circumvents the reducing valve. In others temporary by-passes may be acquired either spontaneously, or as the result of deliberate "spiritual exercises," or through hypnosis, or by means of drugs. Through these permanent or temporary by-passes there flows, not indeed the perception "of everything that is happening everywhere in the universe" (for the by-pass does not abolish the reducing valve, which still excludes the total content of Mind at Large), but something more than, and above all something different from, the carefully selected utilitarian material which our narrowed, individual minds regard as a complete, or at least sufficient, picture of reality.

The brain is provided with a number of enzyme systems which serve to co-ordinate its workings. Some of these enzymes regulate the supply of glucose to the brain cells. Mescalin inhibits the production of these enzymes and thus lowers the amount of glucose available to an organ that is in constant need of sugar. When mescalin reduces the brain's normal ration of sugar what happens? Too few cases have been observed, and therefore a comprehensive answer cannot yet be given. But what happens to the majority of the few who have taken mescalin under supervision can be summarized as follows.

(1) The ability to remember and to "think straight" is little if at all reduced. (Listening to the recordings of my conversation under the influence of the drug, I cannot discover that I was then any stupider than I am at ordinary times.)

(2) Visual impressions are greatly intensified and the eye recovers some of the perceptual innocence of childhood, when the sensum was not immediately and automatically subordinated to the concept. Interest in space is diminished and interest in time falls almost to zero.

(3) Though the intellect remains unimpaired and though perception is enormously improved, the will suffers a profound change for the worse. The mescalin taker sees no reason for doing anything in particular and finds most of the causes for which, at ordinary times, he was prepared to act and suffer, profoundly uninteresting. He can't be bothered with them, for the good reason that he has better things to think about.

(4) These better things may be experienced (as I experienced them) "out there," or "in here," or in both worlds, the inner and the outer, simultaneously or suc-

cessively. That they *are* better seems to be self-evident to all mescalin takers who come to the drug with a sound liver and an untroubled mind.

These effects of mescalin are the sort of effects you could expect to follow the administration of a drug having the power to impair the efficiency of the cerebral reducing valve. When the brain runs out of sugar, the undernourished ego grows weak, can't be bothered to undertake the necessary chores, and loses all interest in those spatial and temporal relationships which mean so much to an organism bent on getting on in the world. As Mind at Large seeps past the no longer watertight valve, all kinds of biologically useless things start to happen. In some cases there may be extra-sensory perceptions. Other persons discover a world of visionary beauty. To others again is revealed the glory, the infinite value and meaningfulness of naked existence, of the given, unconceptualized event. In the final stage of egolessness there is an "obscure knowledge" that All is in all—that All is actually each. This is as near, I take it, as a finite mind can ever come to "perceiving everything that is happening everywhere in the universe."

In this context, how significant is the enormous heightening, under mescalin, of the perception of color! For certain animals it is biologically very important to be able to distinguish certain hues. But beyond the limits of their utilitarian spectrum, most creatures are completely color blind. Bees, for example, spend most of their time "deflowering the fresh virgins of the spring"; but, as Von Frisch has shown, they can recognize only a very few colors. Man's highly developed color sense is a biological luxury—inestimably precious to him as an intellectual and spiritual being, but unnecessary to his survival as an animal. To judge by the adjectives which Homer puts into their mouths, the heroes of the Trojan War hardly excelled the bees in their capacity to distinguish colors. In this respect, at least, mankind's advance has been prodigious.

Mescalin raises all colors to a higher power and makes the percipient aware of innumerable fine shades of difference, to which, at ordinary times, he is completely blind. It would seem that, for Mind at Large, the so-called secondary characters of things are primary. Unlike Locke, it evidently feels that colors are more important, better worth attending to, than masses, positions and dimensions. Like mescalin takers, many mystics perceive supernaturally brilliant colors, not only with the inward eye, but even in the objective world around them. Similar reports are made by psychics and sensitives. There are certain mediums to whom the mescalin taker's brief revelation is a matter, during long periods, of daily and hourly experience.

From this long but indispensable excursion into the realm of theory, we may now return to the miraculous facts—four bamboo chair legs in the middle of a room. Like Wordsworth's daffodils, they brought all manner of wealth—the gift, beyond price, of a new direct insight into the very Nature of Things, together with a more modest treasure of understanding in the field, especially, of the arts.

A rose is a rose is a rose. But these chair legs were chair legs were St. Michael and all angels. Four or five hours after the event, when the effects of a cerebral sugar shortage were wearing off, I was taken for a little tour of the city, which included a visit, towards sundown, to what is modestly claimed to be the World's Biggest Drug Store. At the back of the W.B.D.S., among the toys, the greeting cards and the comics, stood a row, surprisingly enough, of art books. I picked up the first volume that came to hand. It was on Van Gogh, and the picture at which the book opened was "The Chair"—that astounding portrait of a *Ding an Sich*, which the mad painter saw, with a kind of adoring terror, and tried to render on his canvas. But it was a task to which the power even of genius proved wholly inadequate. The chair Van Gogh had seen was obviously the same in essence as the chair I had seen. But, though incomparably more real than the chairs of ordinary perception, the chair in his picture remained no more than an unusually expressive symbol of the fact. The fact had been manifested Suchness; this was only an emblem. Such emblems are sources of true knowledge about the Nature of Things, and this true knowledge may serve to prepare the mind which accepts it for immediate insights on its own account. But that is all. However expressive, symbols can never be the things they stand for.

It would be interesting, in this context, to make a study of the works of art available to the great knowers of Suchness. What sort of pictures did Eckhart look at? What sculptures and paintings played a part in the religious experience of St. John of the Cross, of Hakuin, of Hui-neng, of William Law? The questions are beyond my power to answer; but I strongly suspect that most of the great knowers of Suchness paid very little attention to art—some refusing to have anything to do with it at all, others being content with what a critical eye would regard as second-rate, or even, tenth-rate, works. (To a person whose transfigured and transfiguring mind can see the All in every *this*, the first-rateness or tenth-rateness of even a religious painting will be a matter of the most sovereign indifference.) Art, I suppose, is only for beginners, or else for those resolute dead-enders, who have made up their minds to be content with the *ersatz* of Suchness, with symbols rather than with what they signify, with the elegantly composed recipe in lieu of actual dinner.

I returned the Van Gogh to its rack and picked up the volume standing next to it. It was a book on Botticelli. I turned the pages. "The Birth of Venus"—never one of my favorites. "Mars and Venus," that loveliness so passionately denounced by poor Ruskin at the height of his long-drawn sexual tragedy. The marvelously rich and intricate "Calumny of Apelles." And then a somewhat less familiar and not very good picture, "Judith." My attention was arrested and I gazed in fascination, not at the pale neurotic heroine or her attendant, not at the victim's hairy head or the vernal landscape in the background, but at the purplish silk of Judith's pleated bodice and long wind-blown skirts.

This was something I had seen before—seen that very morning, between the flowers and the furniture, when I looked down by chance, and went on passionately staring by choice, at my own crossed legs. Those folds in the trousers—what a labyrinth of endlessly significant complexity! And the texture of the gray flannel—how rich, how deeply, mysteriously sumptuous! And here they were again, in Botticelli's picture.

Civilized human beings wear clothes, therefore there can be no portraiture, no mythological or historical storytelling without representations of folded textiles. But though it may account for the origins, mere tailoring can never explain the luxuriant development of drapery as a major theme of all the plastic arts. Artists, it is obvious, have always loved drapery for its own sake—or, rather, for their own. When you paint or carve drapery, you are painting or carving forms which, for all practical purposes, are non-representational—the kind of unconditioned forms on which artists even in the most naturalistic tradition like to let themselves go. In the average Madonna or Apostle the strictly human, fully representational element accounts for about ten per cent of the whole. All the rest consists of many colored variations on the inexhaustible theme of crumpled wool or linen. And these non-representational nine-tenths of a Madonna or an Apostle may be just as important qualitatively as they are in quantity. Very often they set the tone of the whole work of art, they state the key in which the theme is being rendered, they express the mood, the temperament, the attitude to life of the artist. Stoical serenity reveals itself in the smooth surfaces, the broad untortured folds of Piero's draperies. Torn between fact and wish, between cynicism and idealism, Bernini tempers the all but caricatural verisimilitude of his faces with enormous sartorial abstractions, which are the embodiment, in stone or bronze, of the everlasting commonplaces of rhetoric—the heroism, the holiness, the sublimity to which mankind perpetually aspires, for the most part in vain. And here are El Greco's disquietingly visceral skirts and mantles; here are the sharp, twisting, flame-like folds in which Cosimo Tura clothes his figures: in the first, traditional spirituality breaks down into a nameless physiological yearning; in the second, there writhes an agonized sense of the world's essential strangeness and hostility. Or consider Watteau; his men and women play lutes, get ready for balls and harlequinades, embark, on velvet lawns and under noble trees, for the Cythera of every lover's dream; their enormous melancholy and the flayed, excruciating sensibility of their creator find expression, not in the actions recorded, not in the gestures and the faces portrayed, but in the relief and texture of their taffeta skirts, their satin capes and doublets. Not an inch of smooth surface here, not a moment of peace or confidence, only a silken wilderness of countless tiny pleats and wrinkles, with an incessant modulation—inner uncertainty rendered with the perfect assurance of a master hand—of tone into tone, of one indeterminate color into another. In life, man proposes, God disposes. In the plastic arts the proposing is done by the subject matter; that which disposes is ultimately the artist's temperament, proximately (at least in portraiture, history and genre) the carved or painted drapery. Between them, these two may decree that a *fête galante* shall move to tears, that a crucifixion shall be serene to the point of cheerfulness, that a stigmatization shall be almost intolerably sexy, that the likeness of a prodigy of female brainlessness (I am thinking now of Ingres' incomparable Mme. Moitessier) shall express the austerest, the most uncompromising intellectuality.

But this is not the whole story. Draperies, as I had now discovered, are much more than devices for the introduction of non-representational forms into naturalistic paintings and sculptures. What the rest of us see only under the influence of mescalin, the artist is congenitally equipped to see all the time. His perception is not limited to what is biologically or socially useful. A little of the knowledge belonging to Mind at Large oozes past the reducing valve of brain and ego, into his consciousness. It is a knowledge of the intrinsic significance of every existent. For the artist as for the mescalin taker draperies are living hieroglyphs that stand in some peculiarly expressive way for the unfathomable mystery of pure being. More even than the chair, though less perhaps than those wholly supernatural flowers, the folds of my gray flannel trousers were charged with "is-ness." To what they owed this privileged status, I cannot say. Is it, perhaps, because the forms of folded drapery are so strange and dramatic that they catch the eye and in this way force the miraculous fact of sheer existence upon the attention? Who knows? What is important is less the reason for the experience than the experience itself. Poring over Judith's skirts, there in the World's Biggest Drug Store, I knew that Botticelli—and not Botticelli alone, but many others too—had looked at draperies with the same transfigured and transfiguring eyes as had been mine that morning. They had seen the *Istigkeit*, the Allness and Infinity of folded cloth and had done their best to render it in paint or stone. Necessarily, of course, without success. For the glory and the wonder of pure existence belong to another order, beyond the power of even the highest art to express. But in Judith's skirt I could clearly see what, if I had been a painter of genius, I might have made of my old gray flannels. Not much, heaven knows, in comparison with the reality, but enough to delight generation after generation of beholders, enough to make them understand at least a little of the true significance of what, in our pathetic imbecility, we call "mere things" and disregard in favor of television.

"This is how one ought to see," I kept saying as I looked down at my trousers, or glanced at the jeweled books in the shelves, at the legs of my infinitely more than Van-Goghian chair. "This is how one ought to see, how things really are." And yet there were reservations. For if one always saw like this, one would never want to do anything else. Just looking, just being the divine Not-self of flower, of book, of chair, of flannel. That would be enough. But in that case what about other people? What about human relations? In the recording of that morning's conversations I find the question constantly repeated, "What about human relations?" How could one reconcile this timeless bliss of seeing as one ought to see with the temporal duties of doing what one ought to do and feeling as one ought to feel? "One ought to be able," I said, "to see these trousers as infinitely important and human beings as still more infinitely important." One ought—but in practice it seemed to be impossible. This participation in the manifest glory of things left no room, so to speak, for the ordinary, the necessary concerns of human existence, above all for concerns involving persons. For persons are selves and, in one respect at least, I was now a Not-self, simultaneously perceiving and being the Not-self of the things around me. To this new-born Not-self, the behavior, the appearance, the very thought of the self it had momentarily ceased to be, and of other selves, its one-time fellows, seemed not indeed distasteful (for distastefulness was not one of the categories in terms of which I was thinking), but enormously irrelevant. Compelled by the investigator to analyze and report on what I was doing (and how I longed to be left alone with Eternity in a

ALDOUS HUXLEY

flower, Infinity in four chair legs and the Absolute in the
folds of a pair of flannel trousers!), I realized that I was
deliberately avoiding the eyes of those who were with me
in the room, deliberately refraining from being too much
aware of them. One was my wife, the other a man I re-
spected and greatly liked; but both belonged to the world
from which, for the moment, mescalin had delivered me
—the world of selves, of time, of moral judgments and
utilitarian considerations, the world (and it was this
aspect of human life which I wished, above all else, to
forget) of self-assertion, of cocksureness, of overvalued
words and idolatrously worshiped notions.

At this stage of the proceedings I was handed a large
colored reproduction of the well-known self-portrait by
Cézanne—the head and shoulders of a man in a large
straw hat, red-cheeked, red-lipped, with rich black
whiskers and a dark unfriendly eye. It is a magnificent
painting; but it was not as a painting that I now saw it.
For the head promptly took on a third dimension and
came to life as a small goblin-like man looking out
through a window in the page before me. I started to
laugh. And when they asked me why, "What pretensions!" I kept repeating. "Who on earth does he think
he is?" The question was not addressed to Cézanne in
particular, but to the human species at large. Who did
they all think they were?

"It's like Arnold Bennett in the Dolomites," I said, suddenly remembering a scene, happily immortalized in a snapshot, of A.B., some four or five years before his death, toddling along a wintry road at Cortina d'Ampezzo. Around him lay the virgin snow; in the background was a more than gothic aspiration of red crags. And there was dear, kind, unhappy A.B., consciously overacting the role of his favorite character in fiction, himself, the Card in person. There he went, toddling slowly in the bright Alpine sunshine, his thumbs in the armholes of a yellow waistcoat which bulged, a little lower down, with the graceful curve of a Regency bow window at Brighton—his head thrown back as though to aim some stammered utterance, howitzer-like, at the blue dome of heaven. What he actually said, I have forgotten; but what his whole manner, air and posture fairly shouted was, "I'm as good as those damned mountains." And in some ways, of course, he was infinitely better; but not, as he knew very well, in the way his favorite character in fiction liked to imagine.

Successfully (whatever that may mean) or unsuccessfully, we all overact the part of our favorite character in fiction. And the fact, the almost infinitely unlikely fact, of actually being Cézanne makes no difference. For the consummate painter, with his little pipeline to Mind at Large by-passing the brain valve and ego-filter, was also and just as genuinely this whiskered goblin with the unfriendly eye.

For relief I turned back to the folds in my trousers. "This is how one ought to see," I repeated yet again. And I might have added, "These are the sort of things one ought to look at." Things without pretensions, satisfied to be merely themselves, sufficient in their Suchness, not acting a part, not trying, insanely, to go it alone, in isolation from the Dharma-Body, in Luciferian defiance of the grace of God.

"The nearest approach to this," I said, "would be a Vermeer."

Yes, a Vermeer. For that mysterious artist was trebly gifted—with the vision that perceives the Dharma-Body as the hedge at the bottom of the garden, with the talent to render as much of that vision as the limitations of human capacity permit, and with the prudence to confine himself in his paintings to the more manageable aspects of reality; for though Vermeer represented human beings, he was always a painter of still life. Cézanne, who told his female sitters to do their best to look like apples, tried to paint portraits in the same spirit. But his pippin-like women are more nearly related to Plato's Ideas than to the Dharma-Body in the hedge. They are Eternity and Infinity seen, not in sand or flower, but in the abstractions of some very superior brand of geometry. Vermeer never asked his girls to look like apples. On the contrary, he insisted on their being girls to the very limit—but always with the proviso that they refrain from behaving girlishly. They might sit or quietly stand but never giggle, never display self-consciousness, never say their prayers or pine for absent sweethearts, never gossip, never gaze enviously at other women's babies, never flirt, never love or hate or work. In the act of doing any of these things they would doubtless become more intensely themselves, but would cease, for that very reason, to manifest their divine essential Not-self. In Blake's phrase, the doors of Vermeer's perception were only partially cleansed. A single panel had become almost perfectly transparent; the rest of the door was still muddy. The essential Not-self could be perceived very clearly in things and in living creatures on the hither side of good and evil. In human beings it was visible only when they were in repose, their minds untroubled, their bodies motionless. In these circumstances Vermeer could see Suchness in all its heavenly beauty—could see and, in some small measure, render it in a subtle and sumptuous still life. Vermeer is undoubtedly the greatest painter of human still lives. But there have been others, for example, Vermeer's French contemporaries, the Le Nain brothers. They set out, I suppose, to be genre painters; but what they actually produced was a series of human still lives, in which their cleansed perception of the infinite significance of all things is rendered not, as with Vermeer, by subtle enrichment of color and texture, but by a heightened clarity, an obsessive distinctness of form, within an austere, almost monochromatic tonality. In our own day we have had Vuillard, the painter, at his best, of unforgettably splendid pictures of the Dharma-Body manifested in a bourgeois bedroom, of the Absolute blazing away in the midst of some stockbroker's family in a suburban garden, taking tea.

> *Ce qui fait que l'ancien bandagiste renie*
> *Le comptoir dont le faste alléchait les passants,*
> *C'est son jardin d'Auteuil, où, veufs de tout encens,*
> *Les Zinnias ont l'air d'être en tôle vernie.*

For Laurent Tailhade the spectacle was merely obscene. But if the retired rubber goods merchant had sat still enough, Vuillard would have seen in him only the Dharma-Body, would have painted, in the zinnias, the goldfish pool, the villa's Moorish tower and Chinese lanterns, a corner of Eden before the Fall.

But meanwhile my question remained unanswered. How was this cleansed perception to be reconciled with a proper concern with human relations, with the necessary chores and duties, to say nothing of charity and practical compassion? The age-old debate between the actives and the contemplatives was being renewed—renewed, so far as I was concerned, with an unprecedented poignancy. For until this morning I had known contemplation only in its humbler, its more ordinary forms—as discursive thinking; as a rapt absorption in poetry or painting or music; as a patient waiting upon those inspirations, without which even the prosiest writer cannot hope to accomplish anything; as occasional glimpses, in Nature, of Wordsworth's "something far more deeply interfused"; as systematic silence leading, sometimes, to hints of an "obscure knowledge." But now I knew contemplation at its height. At its height, but not yet in its fullness. For in its fullness the way of Mary includes the way of Martha and raises it, so to speak, to its own higher power. Mescalin opens up the way of Mary, but shuts the door on that of Martha. It gives access to contemplation—but to a contemplation that is incompatible with action and even with the will to action, the very thought of action. In the intervals between his revelations the mescalin taker is apt to feel that, though in one way everything is supremely as it should be, in another there is something wrong. His problem is essentially the same as that which confronts the quietist, the *arhat* and, on another level, the landscape painter and the painter of human still lives. Mescalin can never solve that problem; it can only pose it, apocalyptically, for those to whom it had never before presented itself. The full and final solution can be found only by those who are prepared to implement the right kind of *Weltanschauung* by means of the right kind of behavior and the right kind of constant and unstrained alertness. Over against the quietist stands the active-contemplative, the saint, the man who, in Eckhart's phrase, is ready to come down from the seventh heaven in order to bring a cup of water to his sick brother. Over against the *arhat*, retreating from appearances into an entirely transcendental Nirvana, stands the Bodhisattva, for whom Suchness and the world of contingencies are one, and for whose boundless compassion every one of those contingencies is an occasion not only for transfiguring insight, but also for the most practical charity. And in the universe of art, over against Vermeer and the other painters of human still lives, over against the masters of Chinese and Japanese landscape painting, over against Constable and Turner, against Sisley and Seurat and Cézanne, stands the all-inclusive art of Rembrandt. These are enormous names, inaccessible eminences. For myself, on this memorable May morning, I could only be grateful for an experience which had shown me, more clearly than I had ever seen it before, the true nature of the challenge and the completely liberating response.

Let me add, before we leave this subject, that there is no form of contemplation, even the most quietistic, which is without its ethical values. Half at least of all morality is negative and consists in keeping out of mischief. The Lord's Prayer is less than fifty words long, and six of those words are devoted to asking God not to lead us into temptation. The one-sided contemplative leaves undone many things that he ought to do; but to make up for it, he refrains from doing a host of things he ought not to do. The sum of evil, Pascal remarked, would be much diminished if men could only learn to sit quietly in their rooms. The contemplative whose perception has been cleansed does not have to stay in his room. He can go about his business, so completely satisfied to see and be a part of the divine Order of Things that he will never even be tempted to indulge in what Traherne called "the dirty Devices of the world." When we feel ourselves to be sole heirs of the universe, when "the sea flows in our veins . . . and the stars are our jewels," when all things are perceived as infinite and holy, what motive can we have for covetousness or self-assertion, for the pursuit of power or the drearier forms of pleasure? Contemplatives are not likely to become gamblers, or procurers, or drunkards; they do not as a rule preach intolerance, or make war; do not find it necessary to rob, swindle or grind the faces of the poor. And to these enormous negative virtues we may add another which, though hard to define, is both positive and important. The *arhat* and the quietist may not practice contemplation in its fullness; but if they practice it at all, they may bring back enlightening reports of another, a transcendent country of the mind; and if they practice it in the height, they will become conduits through which some beneficent influence can flow out of that other country into a world of darkened selves, chronically dying for lack of it.

Meanwhile I had turned, at the investigator's request, from the portrait of Cézanne to what was going on, inside my head, when I shut my eyes. This time, the inscape was curiously unrewarding. The field of vision was filled with brightly colored, constantly changing structures that seemed to be made of plastic or enameled tin.

"Cheap," I commented. "Trivial. Like things in a five-and-ten." And all this shoddiness existed in a closed, cramped universe. "It's as though one were below decks in a ship," I said. "A five-and-ten-cent ship."

And as I looked, it became very clear that this five-and-ten-cent ship was in some way connected with human pretensions, with the portrait of Cézanne, with A.B. among the Dolomites overacting his favorite character in fiction. This suffocating interior of a dime-store ship was my own personal self; these gimcrack mobiles of tin and plastic were my personal contributions to the universe.

I felt the lesson to be salutary, but was sorry, none the less, that it had had to be administered at this moment and in this form. As a rule the mescalin taker discovers an inner world as manifestly a datum, as self-evidently "infinite and holy," as that transfigured outer world which I had seen with my eyes open. From the first, my own case had been different. Mescalin had endowed me temporarily with the power to see things with my eyes shut; but it could not, or at least on this occasion did not, reveal an inscape remotely comparable to my flowers or chair or flannels "out there." What it had allowed me to perceive inside was not the Dharma-Body, in images, but my own mind; not Suchness, but a set of symbols— in other words, a homemade substitute for Suchness.

Most visualizers are transformed by mescalin into visionaries. Some of them—and they are perhaps more numerous than is generally supposed—require no trans-formation; they are visionaries all the time. The mental species to which Blake belonged is fairly widely dis-tributed even in the urban-industrial societies of the present day. The poet-artist's uniqueness does not consist in the fact that (to quote from his *Descriptive Catalogue*) he actually *saw* "those wonderful originals called in the Sacred Scriptures the Cherubim." It does not consist in the fact that "these wonderful originals seen in my visions, were some of them one hundred feet in height . . . all containing mythological and recondite meaning." It consists solely in his ability to render, in words or (somewhat less successfully) in line and color, some hint at least of a not excessively uncommon experience. The untalented visionary may perceive an inner reality no less tremendous, beautiful and significant than the world beheld by Blake; but he lacks altogether the ability to express, in literary or plastic symbols, what he has seen.

From the records of religion and the surviving monuments of poetry and the plastic arts it is very plain that, at most times and in most places, men have attached more importance to the inscape than to objective existents, have felt that what they saw with their eyes shut possessed a spiritually higher significance than what they saw with their eyes open. The reason? Familiarity breeds contempt, and how to survive is a problem ranging in urgency from the chronically tedious to the excruciating. The outer world is what we wake up to every morning of our lives, is the place where, willy-nilly, we must try to make our living. In the inner world there is neither work nor monotony. We visit it only in dreams and musings, and its strangeness is such that we never find the same world on two successive occasions. What wonder, then, if human beings in their search for the divine have generally preferred to look within! Generally, but not always. In their art no less than in their religion, the Taoists and the Zen Buddhists looked beyond visions to the Void, and through the Void at "the ten thousand things" of objective reality. Because of their doctrine of the Word made flesh, Christians should have been able, from the first, to adopt a similar attitude towards the universe around them. But because of the doctrine of the Fall, they found it very hard to do so. As recently as three hundred years ago an expression of thoroughgoing world denial and even world condemnation was both orthodox and comprehensible. "We should feel wonder at nothing at all in Nature except only the Incarnation of Christ." In the seventeenth century, Lallemant's phrase seemed to make sense. Today it has the ring of madness.

In China the rise of landscape painting to the rank of a major art form took place about a thousand, in Japan about six hundred and in Europe about three hundred, years ago. The equation of Dharma-Body with hedge was made by those Zen Masters, who wedded Taoist naturalism with Buddhist transcendentalism. It was, therefore, only in the Far East that landscape painters consciously regarded their art as religious. In the West religious painting was a matter of portraying sacred personages, of illustrating hallowed texts. Landscape painters regarded themselves as secularists. Today we recognize in Seurat one of the supreme masters of what may be called mystical landscape painting. And yet this man who was able, more effectively than any other, to render the One in the many, became quite indignant when somebody praised him for the "poetry" of his work. "I merely apply the System," he protested. In other words he was merely a *pointilliste* and, in his own eyes, nothing else. A similar anecdote is told of John Constable. One day towards the end of his life, Blake met Constable at Hampstead and was shown one of the younger artist's sketches. In spite of his contempt for naturalistic art, the old visionary knew a good thing when he saw it—except, of course, when it was by Rubens. "This is not drawing," he cried, "this is inspiration!" "I had meant it to be drawing," was Constable's characteristic answer. Both men were right. It *was* drawing, precise and veracious, and at the same time it *was* inspiration —inspiration of an order at least as high as Blake's. The pine trees on the Heath had actually been seen as identical with the Dharma-Body. The sketch was a rendering, necessarily imperfect but still profoundly impressive, of what a cleansed perception had revealed to the open eyes of a great painter. From a contemplation, in the tradition of Wordsworth and Whitman, of the Dharma-Body as hedge, and from visions, such as Blake's, of the "wonderful originals" within the mind, contemporary poets have retreated into an investigation of the personal, as opposed to the more than personal, subconscious and to a rendering, in highly abstract terms, not of the given, objective fact, but of mere scientific and theological notions. And something similar has happened in the field of painting, where we have witnessed a general retreat from landscape, the predominant art form of the nineteenth century. This retreat from landscape has not been into that other, inner divine Datum, with which most of the traditional schools of the past were concerned, that Archetypal World, where men have always found the raw materials of myth and religion. No, it has been a retreat from the outward Datum into the personal subconscious, into a mental world more squalid and more tightly closed than even the world of conscious personality. These contraptions of tin and highly colored plastic —where had I seen them before? In every picture gallery that exhibits the latest in nonrepresentational art.

And now someone produced a phonograph and put a record on the turntable. I listened with pleasure, but experienced nothing comparable to my seen apocalypses of flowers or flannel. Would a naturally gifted musician *hear* the revelations which, for me, had been exclusively visual? It would be interesting to make the experiment. Meanwhile, though not transfigured, though retaining its

normal quality and intensity, the music contributed not a
little to my understanding of what had happened to me
and of the wider problems which those happenings had
raised.

Instrumental music, oddly enough, left me rather cold.
Mozart's C-Minor Piano Concerto was interrupted after
the first movement, and a recording of some madrigals
by Gesualdo took its place.

"These voices," I said appreciatively, "these voices—
they're a kind of bridge back to the human world."

And a bridge they remained even while singing the
most startlingly chromatic of the mad prince's compositions. Through the uneven phrases of the madrigals, the
music pursued its course, never sticking to the same key
for two bars together. In Gesualdo, that fantastic character out of a Webster melodrama, psychological disintegration had exaggerated, had pushed to the extreme
limit, a tendency inherent in modal as opposed to fully
tonal music. The resulting works sounded as though they
might have been written by the later Schoenberg.

"And yet," I felt myself constrained to say, as I listened to these strange products of a Counter-Reforma-
tion psychosis working upon a late medieval art form,
"and yet it does not matter that he's all in bits. The whole
is disorganized. But each individual fragment is in order,
is a representative of a Higher Order. The Highest Order

prevails even in the disintegration. The totality is present even in the broken pieces. More clearly present, perhaps, than in a completely coherent work. At least you aren't lulled into a sense of false security by some merely human, merely fabricated order. You have to rely on your immediate perception of the ultimate order. So in a certain sense disintegration may have its advantages. But of course it's dangerous, horribly dangerous. Suppose you couldn't get back, out of the chaos . . ."

From Gesualdo's madrigals we jumped, across a gulf of three centuries, to Alban Berg and the *Lyric Suite*.

"This," I announced in advance, "is going to be hell."

But, as it turned out, I was wrong. Actually the music sounded rather funny. Dredged up from the personal subconscious, agony succeeded twelve-tone agony; but what struck me was only the essential incongruity between a psychological disintegration even completer than Gesualdo's and the prodigious resources, in talent and technique, employed in its expression.

"Isn't he sorry for himself!" I commented with a derisive lack of sympathy. And then, *"Katzenmusik—learned Katzenmusik."* And finally, after a few more minutes of the anguish, "Who cares what his feelings are? Why can't he pay attention to something else?"

As a criticism of what is undoubtedly a very remarkable work, it was unfair and inadequate—but not, I think, irrelevant. I cite it for what it is worth and because that is how, in a state of pure contemplation, I reacted to the *Lyric Suite*.

When it was over, the investigator suggested a walk in the garden. I was willing; and though my body seemed to have dissociated itself almost completely from my mind—or, to be more accurate, though my awareness of the transfigured outer world was no longer accompanied by an awareness of my physical organism—I found myself able to get up, open the French window and walk out with only a minimum of hesitation. It was odd, of course, to feel that "I" was not the same as these arms and legs "out there," as this wholly objective trunk and neck and even head. It was odd; but one soon got used to it. And anyhow the body seemed perfectly well able to look after itself. In reality, of course, it always does look after itself. All that the conscious ego can do is to formulate wishes, which are then carried out by forces which it controls very little and understands not at all. When it does anything more—when it tries too hard, for example, when it worries, when it becomes apprehensive about the future—it lowers the effectiveness of those forces and may even cause the devitalized body to fall ill. In my present state, awareness was not referred to as ego; it was, so to speak, on its own. This meant that the physiological intelligence controlling the body was also on its own. For the moment that interfering neurotic who, in waking hours, tries to run the show, was blessedly out of the way.

From the French window I walked out under a kind of pergola covered in part by a climbing rose tree, in part by laths, one inch wide with half an inch of space between them. The sun was shining and the shadows of the laths made a zebra-like pattern on the ground and across the seat and back of a garden chair, which was standing at this end of the pergola. That chair—shall I ever forget it? Where the shadows fell on the canvas upholstery, stripes of a deep but glowing indigo alternated with stripes of an incandescence so intensely bright that it was hard to believe that they could be made of anything but blue fire. For what seemed an immensely long time I gazed without knowing, even without wishing to know, what it was that confronted me. At any other time I would have seen a chair barred with alternate light and shade. Today the percept had swallowed up the concept. I was so completely absorbed in looking, so thunderstruck by what I actually saw, that I could not be aware of anything else. Garden furniture, laths, sunlight, shadow—these were no more than names and notions, mere verbalizations, for utilitarian or scientific purposes, after the event. The event was this succession of azure furnace doors separated by gulfs of unfathomable gentian. It was inexpressibly wonderful, wonderful to the point, almost, of being terrifying. And suddenly I had an inkling of what it must feel like to be mad. Schizophrenia has its heavens as well as its hells and purgatories. I remember what an old friend, dead these many years, told me about his mad wife. One day in the early stages of the disease, when she still had her lucid intervals he had gone to talk to her about their children. She listened for a time, then cut him short. How could he bear to waste his time on a couple of absent children, when all that really mattered, here and now, was the unspeakable beauty of the patterns he made, in this brown tweed jacket, every time he moved his arms? Alas, this paradise of cleansed perception, of pure one-sided contemplation, was not to endure. The blissful intermissions became rarer, became briefer, until finally there were no more of them; there was only horror.

Most takers of mescalin experience only the heavenly part of schizophrenia. The drug brings hell and purgatory only to those who have had a recent case of jaundice, or who suffer from periodical depressions or a chronic anxiety. If, like the other drugs of remotely comparable power, mescalin were notoriously toxic, the taking of it would be enough, of itself, to cause anxiety. But the reasonably healthy person knows in advance that, so far as he is concerned, mescalin is completely innocuous, that its effects will pass off after eight or ten hours, leaving no hangover and consequently no craving for a renewal of the dose. Fortified by this knowledge, he embarks upon the experiment without fear—in other words, without any disposition to convert an unprecedentedly strange and other than human experience into something appalling, something actually diabolical.

Confronted by a chair which looked like the Last Judgment—or, to be more accurate, by a Last Judgment which, after a long time and with considerable difficulty, I recognized as a chair—I found myself all at once on the brink of panic. This, I suddenly felt, was going too far. Too far, even though the going was into intenser beauty, deeper significance. The fear, as I analyze it in retrospect, was of being overwhelmed, of disintegrating under a pressure of reality greater than a mind, accustomed to living most of the time in a cosy world of symbols, could possibly bear. The literature of religious experience abounds in references to the pains and terrors overwhelming those who have come, too suddenly, face to face with some manifestation of the *Mysterium tremendum*. In theological language, this fear is due to the incompatibility between man's egotism and the divine purity, between man's self-aggravated separateness and the infinity of God. Following Boehme and William Law, we may say that, by unregenerate souls, the divine Light at its full blaze can be apprehended only as a burning, purgatorial fire. An almost identical doctrine is to be found in *The Tibetan Book of the Dead*, where the departed soul is described as shrinking in agony from the Pure Light of the Void, and even from the lesser, tempered Lights, in order to rush headlong into the comforting darkness of selfhood as a reborn human being, or even as a beast, an unhappy ghost, a denizen of hell. Anything rather than the burning brightness of unmitigated Reality—anything!

The schizophrenic is a soul not merely unregenerate, but desperately sick into the bargain. His sickness consists in the inability to take refuge from inner and outer reality (as the sane person habitually does) in the homemade universe of common sense—the strictly human world of useful notions, shared symbols and socially acceptable conventions. The schizophrenic is like a man permanently under the influence of mescalin, and therefore unable to shut off the experience of a reality which he is not holy enough to live with, which he cannot explain away because it is the most stubborn of primary facts, and which, because it never permits him to look at the world with merely human eyes, scares him into interpreting its unremitting strangeness, its burning intensity of significance, as the manifestations of human or even cosmic malevolence, calling for the most desperate countermeasures, from murderous violence at one end of the scale to catatonia, or psychological suicide, at the other. And once embarked upon the downward, the infernal road, one would never be able to stop. That, now, was only too obvious.

"If you started in the wrong way," I said in answer to the investigator's questions, "everything that happened would be a proof of the conspiracy against you. It would all be self-validating. You couldn't draw a breath without knowing it was part of the plot."

"So you think you know where madness lies?"

My answer was a convinced and heartfelt, "Yes."

"And you couldn't control it?"

"No I couldn't control it. If one began with fear and hate as the major premise, one would have to go on to the conclusion."

"Would you be able," my wife asked, "to fix your attention on what *The Tibetan Book of the Dead* calls the Clear Light?"

I was doubtful.

"Would it keep the evil away, if you could hold it? Or would you not be able to hold it?"

I considered the question for some time. "Perhaps," I answered at last, "perhaps I could—but only if there were somebody there to tell me about the Clear Light. One couldn't do it by oneself. That's the point, I suppose, of the Tibetan ritual—someone sitting there all the time and telling you what's what."

After listening to the record of this part of the experiment, I took down my copy of Evans-Wentz's edition of *The Tibetan Book of the Dead,* and opened at random. "O nobly born, let not thy mind be distracted." That was the problem—to remain undistracted. Undistracted by the memory of past sins, by imagined pleasure, by the bitter aftertaste of old wrongs and humiliations, by all the fears and hates and cravings that ordinarily eclipse the Light. What those Buddhist monks did for the dying and the dead, might not the modern psychiatrist do for the insane? Let there be a voice to assure them, by day and even while they are asleep, that in spite of all the terror, all the bewilderment and confusion, the ultimate Reality remains unshakably itself and is of the same substance as the inner light of even the most cruelly tormented mind. By means of such devices as recorders, clock-controlled switches, public address systems and pillow speakers it should be very easy to keep the inmates of even an understaffed institution constantly reminded of this primordial fact. Perhaps a few of the lost souls might in this way be helped to win some measure of control over the universe—at once beautiful and appalling, but always other than human, always totally incomprehensible—in which they find themselves condemned to live.

None too soon, I was steered away from the disquieting splendors of my garden chair. Drooping in green parabolas from the hedge, the ivy fronds shone with a kind of glassy, jade-like radiance. A moment later a clump of Red Hot Pokers, in full bloom, had exploded into my field of vision. So passionately alive that they seemed to be standing on the very brink of utterance, the flowers strained upwards into the blue. Like the chair under the laths, they protected too much. I looked down at the leaves and discovered a cavernous intricacy of the most delicate green lights and shadows, pulsing with undecipherable mystery.

Roses:
The flowers are easy to paint,
The leaves difficult.

Shiki's *haiku* (which I quote in R. H. Blyth's translation) expresses, by indirection, exactly what I then felt— the excessive, the too obvious glory of the flowers, as contrasted with the subtler miracle of their foliage.

We walked out into the street. A large pale blue automobile was standing at the curb. At the sight of it, I was suddenly overcome by enormous merriment. What complacency, what an absurd self-satisfaction beamed from those bulging surfaces of glossiest enamel! Man had created the thing in his own image—or rather in the image of his favorite character in fiction. I laughed till the tears ran down my cheeks.

We re-entered the house. A meal had been prepared. Somebody, who was not yet identical with myself, fell to with ravenous appetite. From a considerable distance and without much interest, I looked on.

When the meal had been eaten, we got into the car and went for a drive. The effects of the mescalin were already on the decline: but the flowers in the gardens still trembled on the brink of being supernatural, the pepper trees and carobs along the side streets still manifestly belonged to some sacred grove. Eden alternated with Dodona. Yggdrasil with the mystic Rose. And then, abruptly, we were at an intersection, waiting to cross Sunset Boulevard. Before us the cars were rolling by in a steady stream—thousands of them, all bright and shiny like an advertiser's dream and each more ludicrous than the last. Once again I was convulsed with laughter.

The Red Sea of traffic parted at last, and we crossed into another oasis of trees and lawns and roses. In a few minutes we had climbed to a vantage point in the hills, and there was the city spread out beneath us. Rather disappointingly, it looked very like the city I had seen on other occasions. So far as I was concerned, transfiguration was proportional to distance. The nearer, the more divinely other. This vast, dim panorama was hardly different from itself.

We drove on, and so long as we remained in the hills, with view succeeding distant view, significance was at its everyday level, well below transfiguration point. The magic began to work again only when we turned down into a new suburb and were gliding between two rows of houses. Here, in spite of the peculiar hideousness of the architecture, there were renewals of transcendental otherness, hints of the morning's heaven. Brick chimneys and green composition roofs glowed in the sunshine, like fragments of the New Jerusalem. And all at once I saw what Guardi had seen and (with what incomparable skill) had so often rendered in his paintings—a stucco wall with a shadow slanting across it, blank but unforgettably beautiful, empty but charged with all the meaning and the mystery of existence. The revelation dawned and was gone again within a fraction of a second. The car had moved on; time was uncovering another manifestation of the eternal Suchness. "Within sameness there is difference. But that difference should be different from sameness is in no wise the intention of all the Buddhas. Their intention is both totality and differentiation." This bank of red and white geraniums, for example—it was entirely different from that stucco wall a hundred yards up the road. But the "is-ness" of both was the same, the eternal quality of their transience was the same.

An hour later, with ten more miles and the visit to the World's Biggest Drug Store safely behind us, we were back at home, and I had returned to that reassuring but profoundly unsatisfactory state known as "being in one's right mind."

That humanity at large will ever be able to dispense with Artificial Paradises seems very unlikely. Most men and women lead lives at the worst so painful, at the best so monotonous, poor and limited that the urge to escape, the longing to transcend themselves if only for a few moments, is and has always been one of the principal appetites of the soul. Art and religion, carnivals and saturnalia, dancing and listening to oratory—all these have served, in H. G. Wells's phrase, as Doors in the Wall. And for private, for everyday use there have always been chemical intoxicants. All the vegetable sedatives and narcotics, all the euphorics that grow on trees, the hallucinogens that ripen in berries or can be squeezed from roots—all, without exception, have been known and systematically used by human beings from time immemorial. And to these natural modifiers of consciousness modern science has added its quota of synthetics—chloral, for example, and benzedrine, the bromides and the barbiturates.

Most of these modifiers of consciousness cannot now be taken except under doctor's orders, or else illegally and at considerable risk. For unrestricted use the West has permitted only alcohol and tobacco. All the other chemical Doors in the Wall are labeled Dope, and their unauthorized takers are Fiends.

We now spend a good deal more on drink and smoke than we spend on education. This, of course, is not surprising. The urge to escape from selfhood and the environment is in almost everyone almost all the time. The urge to do something for the young is strong only in parents, and in them only for the few years during which their children go to school. Equally unsurprising is the current attitude towards drink and smoke. In spite of the growing army of hopeless alcoholics, in spite of the hundreds of thousands of persons annually maimed or killed by drunken drivers, popular comedians still crack jokes about alcohol and its addicts. And in spite of the evidence linking cigarettes with lung cancer, practically everybody regards tobacco smoking as being hardly less normal and natural than eating. From the point of view of the rationalist utilitarian this may seem odd. For the historian, it is exactly what you would expect. A firm conviction of the material reality of Hell never prevented medieval Christians from doing what their ambition, lust or covetousness suggested. Lung cancer, traffic accidents and the millions of miserable and misery-creating alcoholics are facts even more certain than was, in Dante's day, the fact of the Inferno. But all such facts are remote and unsubstantial compared with the near, felt fact of a craving, here and now, for release or sedation, for a drink or a smoke.

Ours is the age, among other things, of the automobile and of rocketing population. Alcohol is incompatible with safety on the roads, and its production, like that of tobacco, condemns to virtual sterility many millions of acres of the most fertile soil. The problems raised by alcohol and tobacco cannot, it goes without saying, be solved by prohibition. The universal and ever-present urge to self-transcendence is not to be abolished by slamming the currently popular Doors in the Wall. The only reasonable policy is to open other, better doors in the hope of inducing men and women to exchange their old bad habits for new and less harmful ones. Some of these other, better doors will be social and technological in nature, others religious or psychological, others dietetic, educational, athletic. But the need for frequent chemical vacations from intolerable selfhood and repulsive surroundings will undoubtedly remain. What is

THE DOORS OF PERCEPTION

needed is a new drug which will relieve and console our suffering species without doing more harm in the long run than it does good in the short. Such a drug must be potent in minute doses and synthesizable. If it does not possess these qualities, its production, like that of wine, beer, spirits and tobacco will interfere with the raising of indispensable food and fibers. It must be less toxic than opium or cocaine, less likely to produce undesirable social consequences than alcohol or the barbiturates, less inimical to heart and lungs than the tars and nicotine of cigarettes. And, on the positive side, it should produce changes in consciousness more interesting, more intrinsically valuable than mere sedation or dreaminess, delusions of omnipotence or release from inhibition.

To most people, mescalin is almost completely innocuous. Unlike alcohol, it does not drive the taker into the kind of uninhibited action which results in brawls, crimes of violence and traffic accidents. A man under the influence of mescalin quietly minds his own business. Moreover, the business he minds is an experience of the most enlightening kind, which does not have to be paid for (and this is surely important) by a compensatory hangover. Of the long-range consequences of regular mescalin taking we know very little. The Indians who consume peyote buttons do not seem to be physically or morally degraded by the habit. However, the available evidence is still scarce and sketchy.*

Although obviously superior to cocaine, opium, alcohol and tobacco, mescalin is not yet the ideal drug. Along with the happily transfigured majority of mescalin takers there is a minority that finds in the drug only hell or purgatory. Moreover, for a drug that is to be used, like alcohol, for general consumption, its effects last for an inconveniently long time. But chemistry and physi-

---

* In his monograph, *Menomini Peyotism*, published (December, 1952) in the Transactions of the American Philosophical Society, Professor J. S. Slotkin has written that "the habitual use of Peyote does not seem to produce any increased tolerance or dependence. I know many people who have been Peyotists for forty to fifty years. The amount of Peyote they use depends upon the solemnity of the occasion; in general they do not take any more Peyote now than they did years ago. Also, there is sometimes an interval of a month or more between rites, and they go without Peyote during this period without feeling any craving for it. Personally, even after a series of rites occurring on four successive weekends, I neither increased the amount of Peyote consumed nor felt any continued need for it." It is evidently with good reason that "Peyote has never been legally declared a narcotic, or its use prohibited by the federal government." However, "during the long history of Indian-white contact, white officials have usually tried to suppress the use of Peyote, because it has been conceived to violate their own mores. But these attempts have always failed." In a footnote Dr. Slotkin adds that "it is amazing to hear the fantastic stories about the effects of Peyote and the nature of the ritual, which are told by the white and Catholic Indian officials in the Menomini Reservation. None of them have had the slightest first-hand experience with the plant or with the religion, yet some fancy themselves to be authorities and write official reports on the subject."

ology are capable nowadays of practically anything. If the psychologists and sociologists will define the ideal, the neurologists and pharmacologists can be relied upon to discover the means whereby that ideal can be realized or at least (for perhaps this kind of ideal can never, in the very nature of things, be fully realized) more nearly approached than in the wine-bibbing past, the whisky-drinking, marijuana-smoking and barbiturate-swallowing present.

The urge to transcend self-conscious selfhood is, as I have said, a principal appetite of the soul. When, for whatever reason, men and women fail to transcend themselves by means of worship, good works and spiritual exercises, they are apt to resort to religion's chemical surrogates—alcohol and "goof pills" in the modern West, alcohol and opium in the East, hashish in the Mohammedan world, alcohol and marijuana in Central America, alcohol and coca in the Andes, alcohol and the barbiturates in the more up-to-date regions of South America. In *Poisons Sacrés, Ivresses Divines* Philippe de Félice has written at length and with a wealth of documentation on the immemorial connection between religion and the taking of drugs. Here, in summary or in direct quotation, are his conclusions. The employment for religious purposes of toxic substances is "extraordinarily widespread. . . . The practices studied in this volume can be observed in every region of the earth, among primitives no less than among those who have reached a high pitch of civilization. We are therefore dealing not with exceptional facts, which might justifiably be overlooked, but with a general and, in the widest sense of the word, a human phenomenon, the kind of phenomenon which cannot be disregarded by anyone who is trying to discover what religion is, and what are the deep needs which it must satisfy."

Ideally, everyone should be able to find self-transcendence in some form of pure or applied religion. In practice it seems very unlikely that this hoped for consummation will ever be realized. There are, and doubtless there always will be, good churchmen and good churchwomen for whom, unfortunately, piety is not enough. The late G. K. Chesterton, who wrote at least as lyrically of drink as of devotion, may serve as their eloquent spokesman.

The modern churches, with some exceptions among the Protestant denominations, tolerate alcohol; but even the most tolerant have made no attempt to convert the drug to Christianity, or to sacramentalize its use. The pious drinker is forced to take his religion in one compartment, his religion-surrogate in another. And perhaps this is inevitable. Drinking cannot be sacramentalized except in religions which set no store on decorum. The worship of Dionysos or the Celtic god of beer was a loud and disorderly affair. The rites of Christianity are incompatible with even religious drunkenness. This does no harm to the distillers, but is very bad for Christianity. Countless persons desire self-transcendence and would be glad to find it in church. But, alas, "the hungry sheep look up and are not fed." They take part in rites, they listen to sermons, they repeat prayers; but their thirst remains unassuaged. Disappointed, they turn to the bottle. For a time at least and in a kind of way, it works. Church may still be attended; but it is no more than the Musical Bank of Butler's *Erewhon*. God may still be acknowledged; but He is God only on the verbal level, only in a strictly Pickwickian sense. The effective object of worship is the bottle and the sole religious experience is that state of uninhibited and belligerent euphoria which follows the ingestion of the third cocktail.

We see, then, that Christianity and alcohol do not and cannot mix. Christianity and mescalin seem to be much more compatible. This has been demonstrated by many tribes of Indians, from Texas to as far north as Wisconsin. Among these tribes are to be found groups affiliated with the Native American Church, a sect whose principal rite is a kind of Early Christian agape, or love feast, where slices of peyote take the place of the sacramental bread and wine. These Native Americans regard the cactus as God's special gift to the Indians, and equate its effects with the workings of the divine Spirit.

Professor J. S. Slotkin, one of the very few white men ever to have participated in the rites of a Peyotist congregation, says of his fellow worshipers that they are "certainly not stupefied or drunk. . . . They never get out of rhythm or fumble their words, as a drunken or stupefied man would do. . . . They are all quiet, courteous and considerate of one another. I have never been in any white man's house of worship where there is either so much religious feeling or decorum." And what, we may ask, are these devout and well-behaved Peyotists experiencing? Not the mild sense of virtue which sustains the average Sunday churchgoer through ninety minutes of boredom. Not even those high feelings, inspired by thoughts of the Creator and the Redeemer, the Judge and the Comforter, which animate the pious. For these Native Americans, religious experience is something more direct and illuminating, more spontaneous, less the homemade product of the superficial, self-conscious mind. Sometimes (according to the reports collected by Dr. Slotkin) they see visions, which may be of Christ Himself. Sometimes they hear the voice of the Great Spirit. Sometimes they become aware of the presence of God and of those personal shortcomings which must be corrected if they are to do His will. The practical consequences of these chemical openings of doors into the Other World seem to be wholly good. Dr. Slotkin reports that habitual Peyotists are on the whole more industrious, more temperate (many of them abstain altogether from alcohol), more peaceable than non-Peyotists. A tree with such satisfactory fruits cannot be condemned out of hand as evil.

In sacramentalizing the use of peyote, the Indians of the Native American Church have done something which is at once psychologically sound and historically respectable. In the early centuries of Christianity many pagan rites and festivals were baptized, so to say, and made to serve the purposes of the Church. These jollifications were not particularly edifying; but they assuaged a certain psychological hunger and, instead of trying to suppress them, the earlier missionaries had the sense to accept them for what they were, soul-satisfying expressions of fundamental urges, and to incorporate them into the fabric of the new religion. What the Native Americans have done is essentially similar. They have taken a pagan custom (a custom, incidentally, far more elevating and enlightening than most of the rather brutish carousals and mummeries adopted from European paganism) and given it a Christian significance.

Though but recently introduced into the northern United States, peyote-eating and the religion based upon it have become important symbols of the red man's right to spiritual independence. Some Indians have reacted to white supremacy by becoming Americanized, others by retreating into traditional Indianism. But some have tried to make the best of both worlds, indeed of all the worlds—the best of Indianism, the best of Christianity, and the best of those Other Worlds of transcendental experience, where the soul knows itself as unconditioned and of like nature with the divine. Hence the Native American Church. In it two great appetites of the soul— the urge to independence and self-determination and the urge to self-transcendence—were fused with, and interpreted in the light of, a third—the urge to worship, to justify the ways of God to man, to explain the universe by means of a coherent theology.

> Lo, the poor Indian, whose untutored mind
> Clothes him in front, but leaves him bare behind.

But actually it is we, the rich and highly educated whites, who have left ourselves bare behind. We cover our anterior nakedness with some philosophy—Christian, Marxian, Freudo-Physicalist—but abaft we remain uncovered, at the mercy of all the winds of circumstance. The poor Indian, on the other hand, has had the wit to protect his rear by supplementing the fig leaf of a theology with the breechclout of transcendental experience.

I am not so foolish as to equate what happens under the influence of mescalin or of any other drug, prepared or in the future preparable, with the realization of the end and ultimate purpose of human life: Enlightenment, the Beatific Vision. All I am suggesting is that the mescalin experience is what Catholic theologians call "a gratuitous grace," not necessary to salvation but potentially helpful and to be accepted thankfully, if made available. To be shaken out of the ruts of ordinary perception, to be shown for a few timeless hours the outer and the inner world, not as they appear to an animal obsessed with survival or to a human being obsessed with words and notions, but as they are apprehended, directly and unconditionally, by Mind at Large—this is an experience of inestimable value to everyone and especially to the intellectual. For the intellectual is by definition the man for whom, in Goethe's phrase, "the word is essentially fruitful." He is the man who feels that "what we perceive by the eye is foreign to us as such and need not impress us deeply." And yet, though himself an intellectual and one of the supreme masters of language, Goethe did not always agree with his own evaluation of the word. "We talk," he wrote in middle life, "far too much. We should talk less and draw more. I personally should like to renounce speech altogether and, like organic Nature, communicate everything I have to say in sketches. That fig tree, this little snake, the cocoon on my window sill quietly awaiting its future—all these are momentous signatures. A person able to decipher their meaning properly would soon be able to dispense with the written or the spoken word altogether. The more I think of it, there is something futile, mediocre, even (I am tempted to say) foppish about speech. By contrast, how the gravity of Nature and her silence startle you, when you stand face to face with her, undistracted, before a barren ridge or in the desolation of the ancient hills." We can never dispense with language and the other symbol systems; for it is by means of them, and only by their means, that we have raised ourselves above the brutes, to the level of human beings. But we can easily become the victims as well as the beneficiaries of these systems. We must learn how to handle words effectively; but at the same time we must preserve and, if necessary, intensify our ability to look at the world directly and not through that half opaque medium of concepts, which distorts every given fact into the all too familiar likeness of some generic label or explanatory abstraction.

Literary or scientific, liberal or specialist, all our education is predominantly verbal and therefore fails to accomplish what it is supposed to do. Instead of transforming children into fully developed adults, it turns out students of the natural sciences who are completely unaware of Nature as the primary fact of experience, it inflicts upon the world students of the humanities who know nothing of humanity, their own or anyone else's.

Gestalt psychologists, such as Samuel Renshaw, have devised methods for widening the range and increasing the acuity of human perceptions. But do our educators apply them? The answer is, No.

Teachers in every field of psycho-physical skill, from seeing to tennis, from tightrope walking to prayer, have discovered, by trial and error, the conditions of optimum functioning within their special fields. But have any of the great Foundations financed a project for co-ordinating these empirical findings into a general theory and practice of heightened creativeness? Again, so far as I am aware, the answer is, No.

All sorts of cultists and queer fish teach all kinds of techniques for achieving health, contentment, peace of mind; and for many of their hearers many of these techniques are demonstrably effective. But do we see respectable psychologists, philosophers and clergymen boldly descending into those odd and sometimes malodorous wells, at the bottom of which poor Truth is so often condemned to sit? Yet once more the answer is, No.

And now look at the history of mescalin research. Seventy years ago men of first-rate ability described the transcendental experiences which come to those who, in good health, under proper conditions and in the right spirit, take the drug. How many philosophers, how many theologians, how many professional educators have had the curiosity to open this Door in the Wall? The answer, for all practical purposes, is, None.

In a world where education is predominantly verbal, highly educated people find it all but impossible to pay serious attention to anything but words and notions. There is always money for, there are always doctorates in, the learned foolery of research into what, for scholars, is the all-important problem: Who influenced whom to say what when? Even in this age of technology the verbal humanities are honored. The non-verbal humanities, the arts of being directly aware of the given facts of our existence, are almost completely ignored. A catalogue, a bibliography, a definitive edition of a third-rate versifier's *ipsissima verba,* a stupendous index to end all indexes—any genuinely Alexandrian project is sure of approval and financial support. But when it comes to finding out how you and I, our children and grandchildren, may become more perceptive, more intensely aware of inward and outward reality, more open to the Spirit, less apt, by psychological malpractices, to make ourselves physically ill, and more capable of controlling our own autonomic nervous system—when it comes to any form of non-verbal education more fundamental (and more likely to be of some practical use) than Swedish drill, no really respectable person in any really respectable university or church will do anything about it. Verbalists are suspicious of the non-verbal; rationalists fear the given, non-rational fact; intellectuals feel that "what we perceive by the eye (or in any other way) is foreign to us as such and need not impress us deeply." Besides, this matter of education in the non-verbal humanities will not fit into any of the established pigeon-holes. It is not religion, not neurology, not gymnastics, not morality or civics, not even experimental psychology. This being so the subject is, for academic and ecclesiastical purposes, non-existent and may safely be ignored altogether or left, with a patronizing smile, to those whom the Pharisees of verbal orthodoxy call cranks, quacks, charlatans and unqualified amateurs.

"I have always found," Blake wrote rather bitterly, "that Angels have the vanity to speak of themselves as the only wise. This they do with a confident insolence sprouting from systematic reasoning."

Systematic reasoning is something we could not, as a species or as individuals, possibly do without. But neither, if we are to remain sane, can we possibly do without direct perception, the more unsystematic the better, of the inner and outer worlds into which we have been born. This given reality is an infinite which passes all understanding and yet admits of being directly and in some sort totally apprehended. It is a transcendence belonging to another order than the human, and yet it may be present to us as a felt immanence, an experienced participation. To be enlightened is to be aware, always, of total reality in its immanent otherness—-to be aware of it and yet to remain in a condition to survive as an animal, to think and feel as a human being, to resort whenever expedient to systematic reasoning. Our goal is to discover that we have always been where we ought to be. Unhappily we make the task exceedingly difficult for ourselves. Meanwhile, however, there are gratuitous graces in the form of partial and fleeting realizations. Under a more realistic, a less exclusively verbal system of education than ours, every Angel (in Blake's sense of that word) would be permitted as a sabbatical treat, would be urged and even, if necessary, compelled to take an occasional trip through some chemical Door in the Wall into the world of transcendental experience. If it terrified him, it would be unfortunate but probably salutary. If it brought him a brief but timeless illumination, so much the better. In either case the Angel might lose a little of the confident insolence sprouting from systematic reasoning and the consciousness of having read all the books.

Near the end of his life Aquinas experienced Infused Contemplation. Thereafter he refused to go back to work on his unfinished book. Compared with *this,* everything he had read and argued about and written—Aristotle and the Sentences, the Questions, the Propositions, the majestic Summas—was no better than chaff or straw. For most intellectuals such a sit-down strike would be inadvisable, even morally wrong. But the Angelic Doctor had done more systematic reasoning than any twelve ordinary Angels, and was already ripe for death. He had earned the right, in those last months of his mortality, to turn away from merely symbolic straw and chaff to the bread of actual and substantial Fact. For Angels of a lower order and with better prospects of longevity, there must be a return to the straw. But the man who comes back through the Door in the Wall will never be quite the same as the man who went out. He will be wiser but less cocksure, happier but less self-satisfied, humbler in acknowledging his ignorance yet better equipped to understand the relationship of words to things, of systematic reasoning to the unfathomable Mystery which it tries, forever vainly, to comprehend.

HEAVEN *and* HELL

In the history of science the collector of specimens preceded the zoologist and followed the exponents of natural theology and magic. He had ceased to study animals in the spirit of the authors of the bestiaries, for whom the ant was incarnate industry, the panther an emblem, surprisingly enough, of Christ, the polecat a shocking example of uninhibited lasciviousness. But, except in a rudimentary way, he was not yet a physiologist, ecologist or student of animal behavior. His primary concern was to make a census, to catch, kill, stuff and describe as many kinds of beasts as he could lay his hands on.

Like the earth of a hundred years ago, our mind still has its darkest Africas, its unmapped Borneos and Amazonian basins. In relation to the fauna of these regions we are not yet zoologists, we are mere naturalists and collectors of specimens. The fact is unfortunate; but we have to accept it, we have to make the best of it. However lowly, the work of the collector must be done, before we can proceed to the higher scientific tasks of classification, analysis, experiment and theory making.

Like the giraffe and the duckbilled platypus, the creatures inhabiting these remoter regions of the mind are exceedingly improbable. Nevertheless they exist, they are facts of observation; and as such, they cannot be ignored by anyone who is honestly trying to understand the world in which he lives.

It is difficult, it is all but impossible, to speak of mental events except in similes drawn from the more familiar universe of material things. If I have made use of geographical and zoological metaphors, it is not wantonly, out of a mere addiction to picturesque language. It is because such metaphors express very forcibly the essential otherness of the mind's far continents, the complete autonomy and self-sufficiency of their inhabitants. A man consists of what I may call an Old World of personal consciousness and, beyond a dividing sea, a series of New Worlds—the not too distant Virginias and Carolinas of the personal subconscious and the vegetative soul; the Far West of the collective unconscious, with its flora of symbols, its tribes of aboriginal archetypes; and, across another, vaster ocean, at the antipodes of everyday consciousness, the world of Visionary Experience.

If you go to New South Wales, you will see marsupials hopping about the countryside. And if you go to the antipodes of the self-conscious mind, you will encounter all sorts of creatures at least as odd as kangaroos. You do not invent these creatures any more than you invent marsupials. They live their own lives in complete independence. A man cannot control them. All he can do is to go to the mental equivalent of Australia and look around him.

Some people never consciously discover their antipodes. Others make an occasional landing. Yet others (but they are few) find it easy to go and come as they please. For the naturalist of the mind, the collector of psychological specimens, the primary need is some safe, easy and reliable method of transporting himself and others from the Old World to the New, from the continent of familiar cows and horses to the continent of a wallaby and the platypus.

Two such methods exist. Neither of them is perfect; but both are sufficiently reliable, sufficiently easy and sufficiently safe to justify their employment by those who know what they are doing. In the first case the soul is transported to its far-off destination by the aid of a chemical—either mescalin or lysergic acid. In the second case, the vehicle is psychological in nature, and the passage to the mind's antipodes is accomplished by hypnosis. The two vehicles carry the consciousness to the same region; but the drug has the longer range and takes its passengers further into the *terra incognita.**

How and why does hypnosis produce its observed effects? We do not know. For our present purposes, however, we do not have to know. All that is necessary, in this context, is to record the fact that some hypnotic subjects are transported, in the trance state, to a region in the mind's antipodes, where they find the equivalent of marsupials—strange psychological creatures leading an autonomous existence according to the law of their own being.

About the physiological effects of mescalin we know a little. Probably (for we are not yet certain) it interferes with the enzyme system that regulates cerebral functioning. By doing so it lowers the efficiency of the brain as an instrument for focusing the mind on the problems of life on the surface of our planet. This lowering of what may be called the biological efficiency of the brain seems to

* See Appendix I, page 143.

permit the entry into consciousness of certain classes of mental events, which are normally excluded, because they possess no survival value. Similar intrusions of biologically useless, but aesthetically and sometimes spiritually valuable material may occur as the result of illness or fatigue; or they may be induced by fasting, or a period of confinement in a place of darkness and complete silence.*

A person under the influence of mescalin or lysergic acid will stop seeing visions when given a large dose of nicotinic acid. This helps to explain the effectiveness of fasting as an inducer of visionary experience. By reducing the amount of available sugar, fasting lowers the brain's biological efficiency and so makes possible the entry into consciousness of material possessing no survival value. Moreover, by causing a vitamin deficiency, it removes from the blood that known inhibitor of visions, nicotinic acid. Another inhibitor of visionary experience is ordinary, everyday, perceptual experience. Experimental psychologists have found that, if you confine a man to a "restricted environment," where there is no light, no sound, nothing to smell and, if you put him in a tepid bath, only one, almost imperceptible thing to touch,

*See Appendix II, page 149.

ALDOUS HUXLEY

the victim will very soon start "seeing things," "hearing things" and having strange bodily sensations.

Milarepa, in his Himalayan cavern, and the anchorites of the Thebaid followed essentially the same procedure and got essentially the same results. A thousand pictures of the Temptations of St. Anthony bear witness to the effectiveness of restricted diet and restricted environment. Asceticism, it is evident, has a double motivation. If men and women torment their bodies, it is not only because they hope in this way to atone for past sins and avoid future punishments; it is also because they long to visit the mind's antipodes and do some visionary sightseeing. Empirically and from the reports of other ascetics, they know that fasting and a restricted environment will transport them where they long to go. Their self-inflicted punishment may be the door to paradise. (It may also—and this is a point which will be discussed in a later paragraph —be a door into the infernal regions.)

From the point of view of an inhabitant of the Old World, marsupials are exceedingly odd. But oddity is not the same as randomness. Kangaroos and wallabies may lack verisimilitude; but their improbability repeats itself and obeys recognizable laws. The same is true of the psychological creatures inhabiting the remoter regions of our minds. The experiences encountered under the influence of mescalin or deep hypnosis are very strange; but they are strange with a certain regularity, strange according to a pattern.

What are the common features which this pattern imposes upon our visionary experiences? First and most important is the experience of light. Everything seen by those who visit the mind's antipodes is brilliantly illuminated and seems to shine from within. All colors are intensified to a pitch far beyond anything seen in the normal state, and at the same time the mind's capacity for recognizing fine distinctions of tone and hue is notably heightened.

In this respect there is a marked difference between these visionary experiences and ordinary dreams. Most dreams are without color, or else are only partially or feebly colored. On the other hand, the visions met with under the influence of mescalin or hypnosis are always intensely and, one might say, preternaturally brilliant in color. Professor Calvin Hall, who has collected records of many thousands of dreams, tells us that about two-thirds of all dreams are in black and white. "Only one dream in three is colored, or has some color in it." A few people dream entirely in color; a few never experience color in their dreams; the majority sometimes dream in color, but more often do not.

"We have come to the conclusion," writes Dr. Hall, "that color in dreams yields no information about the personality of the dreamer." I agree with this conclusion. Color in dreams and visions tells us no more about the personality of the beholder than does color in the external world. A garden in July is perceived as brightly colored. The perception tells us something about sunshine, flowers and butterflies, but little or nothing about our own selves. In the same way the fact that we see brilliant colors in our visions and in some of our dreams tells us something about the fauna of the mind's antipodes, but nothing whatever about the personality who inhabits what I have called the Old World of the mind.

Most dreams are concerned with the dreamer's private wishes and instinctive urges, and with the conflicts which arise when these wishes and urges are thwarted by a disapproving conscience or a fear of public opinion. The story of these drives and conflicts is told in terms of dramatic symbols, and in most dreams the symbols are uncolored. Why should this be the case? The answer, I presume, is that, to be effective, symbols do not require to be colored. The letters in which we write about roses need not be red, and we can describe the rainbow by means of ink marks on white paper. Textbooks are illustrated by line engravings and half-tone plates; and these uncolored images and diagrams effectively convey information.

What is good enough for the waking consciousness is evidently good enough for the personal subconscious, which finds it possible to express its meanings through uncolored symbols. Color turns out to be a kind of touchstone of reality. That which is given is colored; that which our symbol-creating intellect and fancy put together is uncolored. Thus the external world is perceived as colored. Dreams, which are not given but fabricated by the personal subconscious, are generally in black and white. (It is worth remarking that, in most people's experience, the most brightly colored dreams are those of landscapes, in which there is no drama, no symbolic reference to conflict, merely the presentation to consciousness of a given, non-human fact.)

The images of the archetypal world are symbolic; but since we, as individuals, do not fabricate them, but find them "out there" in the collective unconscious, they exhibit some at least of the characteristics of given reality and are colored. The non-symbolic inhabitants of the

ALDOUS HUXLEY

mind's antipodes exist in their own right, and like the
given facts of the external world are colored. Indeed, they
are far more intensely colored than external data. This
may be explained, at least in part, by the fact that our
perceptions of the external world are habitually clouded
by the verbal notions in terms of which we do our thinking. We are forever attempting to convert things into
signs for the more intelligible abstractions of our own
invention. But in doing so, we rob these things of a great
deal of their native thinghood.

At the antipodes of the mind, we are more or less completely free of language, outside the system of conceptual
thought. Consequently our perception of visionary objects possesses all the freshness, all the naked intensity, of
experiences which have never been verbalized, never
assimilated to lifeless abstractions. Their color (that hall-
mark of givenness) shines forth with a brilliance which
seems to us preternatural, because it is in fact entirely
natural—entirely natural in the sense of being entirely
unsophisticated by language or the scientific, philosophical and utilitarian notions, by means of which we ordinarily re-create the given world in our own drearily
human image.

In his *Candle of Vision*, the Irish poet, Æ (George

Russell), has analyzed his visionary experiences with re-markable acuity. "When I meditate," he writes, "I feel in the thoughts and images that throng about me the reflections of personality; but there are also windows in the soul, through which can be seen images created not by human but by the divine imagination."

Our linguistic habits lead us into error. For example, we are apt to say, "I imagine," when what we should have said is, "The curtain was lifted that I might see." Spontaneous or induced, visions are never our personal property. Memories belonging to the ordinary self have no place in them. The things seen are wholly unfamiliar. "There is no reference or resemblance," in Sir William Herschel's phrase, "to any objects recently seen or even thought of." When faces appear, they are never the faces of friends or acquaintances. We are out of the Old World, and exploring the antipodes.

For most of us most of the time, the world of everyday experience seems rather dim and drab. But for a few people often, and for a fair number occasionally, some of the brightness of visionary experience spills over, as it were, into common seeing, and the everyday universe is transfigured. Though still recognizably itself, the Old World takes on the quality of the mind's antipodes. Here is an entirely characteristic description of this transfiguration of the everyday world:

"I was sitting on the seashore, half listening to a friend arguing violently about something which merely bored me. Unconsciously to myself, I looked at a film of sand I had picked up on my hand, when I suddenly saw the exquisite beauty of every little grain of it; instead of being dull, I saw that each particle was made up on a perfect geometrical pattern, with sharp angles, from each of which a brilliant shaft of light was reflected, while each tiny crystal shone like a rainbow. . . . The rays crossed and recrossed, making exquisite patterns of such beauty that they left me breathless. . . . Then, suddenly, my consciousness was lighted up from within and I saw in a vivid way how the whole universe was made up of particles of material which, no matter how dull and lifeless they might seem, were nevertheless filled with this intense and vital beauty. For a second or two the whole world appeared as a blaze of glory. When it died down, it left me with something I have never forgotten and which constantly reminds me of the beauty locked up in every minute speck of material around us."

Similarly George Russell writes of seeing the world illumined by "an intolerable lustre of light"; of finding himself looking at "landscapes as lovely as a lost Eden"; of beholding a world where the "colors were brighter and purer, and yet made a softer harmony." Again, "the winds were sparkling and diamond clear, and yet full of color as an opal, as they glittered through the valley, and I knew the Golden Age was all about me, and it was we who had been blind to it, but that it had never passed away from the world."

Many similar descriptions are to be found in the poets and in the literature of religious mysticism. One thinks, for example, of Wordsworth's *Ode on the Intimations of Immortality in Childhood;* of certain lyrics by George Herbert and Henry Vaughan; of Traherne's *Centuries of Meditation;* of the passage in his autobiography, where Father Surin describes the miraculous transformation of an enclosed convent garden into a fragment of heaven.

Preternatural light and color are common to all visionary experiences. And along with light and color there goes, in every case, a recognition of heightened significance. The self-luminous objects which we see in the mind's antipodes possess a meaning, and this meaning is, in some sort, as intense as their color. Significance here is identical with being; for, at the mind's antipodes, objects do not stand for anything but themselves. The images which appear in the nearer reaches of the collective sub-conscious have meaning in relation to the basic facts of human experience; but here, at the limits of the visionary world, we are confronted by facts which, like the facts of external nature, are independent of man, both individually and collectively, and exist in their own right. And their meaning consists precisely in this, that they are intensely themselves and, being intensely themselves, are manifestations of the essential givenness, the non-human otherness of the universe.

Light, color and significance do not exist in isolation. They modify, or are manifested by, objects. Are there any special classes of objects common to most visionary experiences? The answer is: yes, there are. Under mescalin and hypnosis, as well as in spontaneous visions, certain classes of perceptual experiences turn up again and again.

The typical mescalin or lysergic-acid experience begins with perceptions of colored, moving, living geometrical forms. In time, pure geometry becomes concrete, and the visionary perceives, not patterns, but patterned things, such as carpets, carvings, mosaics. These give place to vast and complicated buildings, in the midst of landscapes, which change continuously, passing from richness to more intensely colored richness, from grandeur to deepening grandeur. Heroic figures, of the kind that Blake called "The Seraphim," may make their appearance, alone or in multitudes. Fabulous animals move across the scene. Everything is novel and amazing. Almost never does the visionary see anything that reminds him of his own past. He is not remembering scenes, persons or objects, and he is not inventing them; he is looking on at a new creation.

The raw material for this creation is provided by the visual experiences of ordinary life; but the molding of this material into forms is the work of someone who is most certainly not the self, who originally had the experiences, or who later recalled and reflected upon them. They are (to quote the words used by Dr. J. R. Smythies in a recent paper in the *American Journal of Psychiatry*) "the work of a highly differentiated mental compartment, without any apparent connection, emotional or volitional, with the aims, interests, or feelings of the person concerned."

Here, in quotation or condensed paraphrase, is Weir Mitchell's account of the visionary world to which he was transported by peyote, the cactus which is the natural source of mescalin.

At his entry into that world he saw a host of "star points" and what looked like "fragments of stained glass." Then came "delicate floating films of color." These were displaced by an "abrupt rush of countless points of white light," sweeping across the field of vision. Next there were zigzag lines of very bright colors, which somehow turned into swelling clouds of still more brilliant hues. Buildings now made their appearance, and then landscapes. There was a Gothic tower of elaborate design with worn statues in the doorways or on stone brackets. "As I gazed, every projecting angle, cornice and even the faces of the stones at their joinings were by degrees covered or hung with clusters of what seemed to be huge precious stones, but uncut stones, some being more like masses of transparent fruit. . . . All seemed to possess an interior light." The Gothic tower gave place to a mountain, a cliff of inconceivable height, a colossal bird claw carved in stone and projecting over the abyss, an endless unfurling of colored draperies, and an efflorescence of more precious stones. Finally there was a view of green and purple waves breaking on a beach "with myriads of lights of the same tint as the waves."

Every mescalin experience, every vision arising under hypnosis, is unique; but all recognizably belong to the same species. The landscapes, the architectures, the clustering gems, the brilliant and intricate patterns—these, in their atmosphere of preternatural light, preternatural color and preternatural significance, are the stuff of which the mind's antipodes are made. Why this should be so, we have no idea. It is a brute fact of experience which, whether we like it or not, we have to accept—just as we have to accept the fact of kangaroos.

From these facts of visionary experience let us now pass to the accounts preserved in all the cultural traditions, of Other Worlds—the worlds inhabited by the gods, by the spirits of the dead, by man in his primal state of innocence.

Reading these accounts, we are immediately struck by the close similarity between induced or spontaneous visionary experience and the heavens and fairylands of folklore and religion. Preternatural light, preternatural intensity of coloring, preternatural significance—these are characteristic of all the Other Worlds and Golden Ages. And in virtually every case this preternaturally significant light shines on, or shines out of, a landscape of such surpassing beauty that words cannot express it.

Thus in the Greco-Roman tradition we find the lovely Garden of the Hesperides, the Elysian Plain, and the fair

Island of Leuke, to which Achilles was translated. Memnon went to another luminous island, somewhere in the East. Odysseus and Penelope traveled in the opposite direction and enjoyed their immortality with Circe in Italy. Still further to the west were the Islands of the Blest, first mentioned by Hesiod and so firmly believed in that, as late as the first century B.C., Sertorius planned to send a squadron from Spain to discover them.

Magically lovely islands reappear in the folklore of the Celts and, at the opposite side of the world, in that of the Japanese. And between Avalon in the extreme West and Horaisan in the Far East, there is the land of Uttarakuru, the Other World of the Hindus. "The land," we read in the *Ramayana,* "is watered by lakes with golden lotuses. There are rivers by thousands, full of leaves of the color of sapphire and lapis lazuli; and the lakes, resplendent like the morning sun, are adorned by golden beds of red lotus. The country all around is covered by jewels and precious stones, with gay beds of blue lotus, golden-petalled. Instead of sand, pearls, gems and gold form the banks of the rivers, which are overhung with trees of fire-bright gold. These trees perpetually bear flowers and fruit, give forth a sweet fragrance and abound with birds."

Uttarakuru, we see, resembles the landscapes of the

mescalin experience in being rich with precious stones. And this characteristic is common to virtually all the Other Worlds of religious tradition. Every paradise abounds in gems, or at least in gemlike objects resembling, as Weir Mitchell puts it, "transparent fruit." Here, for example, is Ezekiel's version of the Garden of Eden. "Thou hast been in Eden, the garden of God. Every precious stone was thy covering, the sardius, topaz and the diamond, the beryl, the onyx and the jasper, the sapphire, the emerald and the carbuncle, and gold. . . . Thou art the anointed cherub that covereth . . . thou hast walked up and down in the midst of the stones of fire." The Buddhist paradises are adorned with similar "stones of fire." Thus, the Western Paradise of the Pure Land Sect is walled with silver, gold and beryl; has lakes with jeweled banks and a profusion of glowing lotuses, within which the bodhisattvas sit enthroned.

In describing their Other Worlds, the Celts and Teutons speak very little of precious stones, but have much to say of another and, for them, equally wonderful substance—glass. The Welsh had a blessed land called Ynisvitrin, the Isle of Glass; and one of the names of the Germanic kingdom of the dead was Glasberg. One is reminded of the Sea of Glass in the Apocalypse.

Most paradises are adorned with buildings, and, like the trees, the waters, the hills and fields, these buildings are bright with gems. We are familiar with the New Jerusalem. "And the building of the wall of it was of jasper, and the city was of pure gold, like unto clear glass. . . . And the foundations of the wall of the city were garnished with all manner of precious stones."

Similar descriptions are to be found in the eschatological literature of Hinduism, Buddhism and Islam. Heaven is always a place of gems. Why should this be the case? Those who think of all human activities in terms of a social and economic frame of reference will give some such answer as this: Gems are very rare on earth. Few people possess them. To compensate themselves for these facts, the spokesmen for the poverty-stricken majority have filled their imaginary heavens with precious stones. This "pie in the sky" hypothesis contains, no doubt, some element of truth; but it fails to explain why precious stones should have come to be regarded as precious in the first place.

Men have spent enormous amounts of time, energy and money on the finding, mining and cutting of colored pebbles. Why? The utilitarian can offer no explanation for such fantastic behavior. But as soon as we take into account the facts of visionary experience, everything becomes clear. In vision, men perceive a profusion of what Ezekiel calls "stones of fire," of what Weir Mitchell describes as "transparent fruit." These things are self-luminous, exhibit a preternatural brilliance of color and possess a preternatural significance. The material objects which most nearly resemble these sources of visionary illumination are gem stones. To acquire such a stone is to acquire something whose preciousness is guaranteed by the fact that it exists in the Other World.

Hence man's otherwise inexplicable passion for gems and hence his attribution to precious stones of therapeutic and magical virtue. The causal chain, I am convinced, begins in the psychological Other World of visionary experience, descends to earth and mounts again to the theological Other World of heaven. In this context the words of Socrates, in the *Phaedo*, take on a new significance. There exists, he tells us, an ideal world above and beyond the world of matter. "In this other earth the colors are much purer and much more brilliant than they are down here. . . . The very mountains, the very stones have a richer gloss, a lovelier transparency and intensity of hue. The precious stones of this lower world, our highly prized cornelians, jaspers, emeralds and all the rest, are but the tiny fragments of these stones above. In the other earth there is no stone but is precious and exceeds in beauty every gem of ours."

In other words, precious stones are precious because they bear a faint resemblance to the glowing marvels seen with the inner eye of the visionary. "The view of that world," says Plato, "is a vision of blessed beholders"; for to see things "as they are in themselves" is bliss unalloyed and inexpressible.

Among people who have no knowledge of precious stones or of glass, heaven is adorned not with minerals, but flowers. Preternaturally brilliant flowers bloom in most of the Other Worlds described by primitive eschatologists, and even in the begemmed and glassy paradises of the more advanced religions they have their place. One remembers the lotus of Hindu and Buddhist tradition, the roses and lilies of the West.

"God first planted a garden." The statement expresses a deep psychological truth. Horticulture has its source—or at any rate one of its sources—in the Other World of the mind's antipodes. When worshipers offer flowers at the altar, they are returning to the gods things which they know, or (if they are not visionaries) obscurely feel, to be indigenous to heaven.

And this return to the source is not merely symbolical; it is also a matter of immediate experience. For the traffic between our Old World and its antipodes, between Here and Beyond, travels along a two-way street. Gems, for example, come from the soul's visionary heaven; but they also lead the soul back to that heaven. Contemplating them, men find themselves (as the phrase goes) *transported*—carried away toward that Other Earth of the Platonic dialogue, that magical place where every pebble is a precious stone. And the same effects may be produced by artifacts of glass and metal, by tapers burning in the dark, by brilliantly colored images and ornaments; by flowers, shells and feathers; by landscapes seen, as Shelley from the Euganean Hills saw Venice, in the transfiguring light of dawn or sunset.

Indeed, we may risk a generalization and say that whatever, in nature or in a work of art, resembles one of those intensely significant, inwardly glowing objects encountered at the mind's antipodes is capable of inducing, if only in a partial and attenuated form, the visionary experience. At this point a hypnotist will remind us that, if he can be induced to stare intently at a shiny object, a patient may go into trance; and that if he goes into

trance, or if he goes only into reverie, he may very well
see visions within and a transfigured world without.

But how, precisely, and why does the view of a shiny
object induce a trance or a state of reverie? Is it, as the
Victorians maintained, a simple matter of eye strain resulting in general nervous exhaustion? Or shall we explain the phenomenon in purely psychological terms—as
concentration pushed to the point of mono-ideism and
leading to dissociation?

And there is a third possibility. Shiny objects may
remind our unconscious of what it enjoys at the mind's
antipodes, and these obscure intimations of life in the
Other World are so fascinating that we pay less attention to this world and so become capable of experiencing
consciously something of that which, unconsciously, is
always with us.

We see then that there are in nature certain scenes,
certain classes of objects, certain materials, possessed of
the power to transport the beholder's mind in the direction of its antipodes, out of the everyday Here and toward
the Other World of Vision. Similarly, in the realm of art,
we find certain works, even certain classes of works, in
which the same transporting power is manifest. These
vision-inducing works may be executed in vision-induc-

ing materials, such as glass, metal, gems or gemlike pigments. In other cases their power is due to the fact that they render, in some peculiarly expressive way, some transporting scene or object.

The best vision-inducing art is produced by men and women who have themselves had the visionary experience; but it is also possible for any reasonably good artist, simply by following an approved recipe, to create works which shall have at least some transporting power.

Of all the vision-inducing arts that which depends most completely on its raw materials is, of course, the art of the goldsmith and jeweler. Polished metals and precious stones are so intrinsically transporting that even a Victorian, even an *art nouveau* jewel is a thing of power. And when to this natural magic of glinting metal and self-luminous stone is added the other magic of noble forms and colors artfully blended, we find ourselves in the presence of a genuine talisman.

Religious art has always and everywhere made use of these vision-inducing materials. The shrine of gold, the chryselephantine statue, the jeweled symbol or image, the glittering furniture of the altar—we find these things in contemporary Europe as in ancient Egypt, in India and China as among the Greeks, the Incas, the Aztecs.

The products of the goldsmith's art are intrinsically numinous. They have their place at the very heart of every Mystery, in every holy of holies. This sacred jewelry has always been associated with the light of lamps and candles. For Ezekiel, a gem was a stone of fire. Conversely, a flame is a living gem, endowed with all the transporting power that belongs to the precious stone and, to a lesser degree, to polished metal. This transporting power of flame increases in proportion to the depth and extent of the surrounding darkness. The most impressively numinous temples are caverns of twilight, in which a few tapers give life to the transporting, other-worldly treasures on the altar.

Glass is hardly less effective as an inducer of visions than are the natural gems. In certain respects, indeed, it is more effective, for the simple reason that there is more of it. Thanks to glass, a whole building—the Sainte Chapelle, for example, the cathedrals of Chartres and Sens—could be turned into something magical and transporting. Thanks to glass, Paolo Uccello could design a circular jewel thirteen feet in diameter—his great window of the Resurrection, perhaps the most extraordinary single work of vision-inducing art ever produced.

For the men of the Middle Ages, it is evident, visionary

experience was supremely valuable. So valuable, indeed, that they were ready to pay for it in hard-earned cash. In the twelfth century collecting boxes were placed in the churches for the upkeep and installation of stained-glass windows. Suger, the Abbot of St. Denis, tells us that they were always full.

But self-respecting artists cannot be expected to go on doing what their fathers have already done supremely well. In the fourteenth century color gave place to grisaille, and windows ceased to be vision inducing. When, in the later fifteenth century, color came into fashion again, the glass painters felt the desire, and found themselves, at the same time, technically equipped, to imitate Renaissance painting in transparency. The results were often interesting; but they were not transporting.

Then came the Reformation. The Protestants disapproved of visionary experience and attributed a magical virtue to the printed word. In a church with clear windows the worshipers could read their Bibles and prayer books and were not tempted to escape from the sermon into the Other World. On the Catholic side the men of the Counter Reformation found themselves in two minds. They thought visionary experience was a good thing, but they also believed in the supreme value of print.

In the new churches stained glass was rarely installed, and in many of the older churches it was wholly or partially replaced by clear glass. The unobscured light permitted the faithful to follow the service in their books, and at the same time to see the vision-inducing works created by the new generations of baroque sculptors and architects. These transporting works were executed in metal and polished stone. Wherever the worshiper turned, he found the glint of bronze, the rich radiance of colored marble, the unearthly whiteness of statuary.

On the rare occasions when the Counter Reformers made use of glass, it was as a surrogate for diamonds, not for rubies or sapphires. Faceted prisms entered religious art in the seventeenth century, and in Catholic churches they dangle to this day from innumerable chandeliers. (These charming and slightly ridiculous ornaments are among the very few vision-inducing devices permitted in Islam. Mosques have no images or reliquaries; but in the Near East, at any rate, their austerity is sometimes mitigated by the transporting glitter of rococo crystal.)

From glass, stained or cut, we pass to marble and the other stones that take a high polish and can be used in mass. The fascination exercised by such stones may be gauged by the amount of time and trouble spent in obtaining them. At Baalbek, for example, and, two or three hundred miles further inland, at Palmyra, we find among the ruins columns of pink granite from Aswan. These great monoliths were quarried in Upper Egypt, were floated in barges down the Nile, were towed across the Mediterranean to Byblos or Tripolis and from thence were hauled, by oxen, mules and men, uphill to Homs, and from Homs southward to Baalbek, or east, across the desert, to Palmyra.

What a labor of giants! And, from the utilitarian point of view, how marvelously pointless! But in fact, of course, there was a point—a point that existed in a region beyond mere utility. Polished to a visionary glow, the rosy shafts proclaimed their manifest kinship with the Other World. At the cost of enormous efforts men had transported these stones from their quarry on the Tropic of Cancer; and now, by way of recompense, the stones were transporting their transporters halfway to the mind's visionary antipodes.

The question of utility and of the motives that lie beyond utility arises once more in relation to ceramics. Few things are more useful, more absolutely indispensable, than pots and plates and jugs. But at the same time few human beings concern themselves less with utility than do the collectors of porcelain and glazed earthenware. To say that these people have an appetite for beauty is not a sufficient explanation. The commonplace ugliness of the surroundings, in which fine ceramics are so often displayed, is proof enough that what their owners crave is not beauty in all its manifestations, but only a special kind of beauty—the beauty of curved reflections, of softly lustrous glazes, of sleek and smooth surfaces. In a word, the beauty that transports the beholder, because it reminds him, obscurely or explicitly, of the preternatural lights and colors of the Other World. In the main the art of the potter has been a secular art—but a secular art which its innumerable devotees have treated with an almost idolatrous reverence. From time to time, however, this secular art has been placed at the service of religion. Glazed tiles have found their way into mosques and, here and there, into Christian churches. From China come shining ceramic images of gods and saints. In Italy Luca della Robbia created a heaven of blue glaze, for his lustrous white madonnas and Christ children. Baked clay is cheaper than marble and, suitably treated, almost as transporting.

Plato and, during a later flowering of religious art, St.

Thomas Aquinas maintained that pure, bright colors were of the very essence of artistic beauty. A Matisse, in that case, would be intrinsically superior to a Goya or a Rembrandt. One has only to translate the philosophers' abstractions into concrete terms to see that this equation of beauty in general with bright, pure colors is absurd. But though untenable as it stands, the venerable doctrine is not altogether devoid of truth.

Bright, pure colors are characteristic of the Other World. Consequently works of art painted in bright, pure colors are capable, in suitable circumstances, of transporting the beholder's mind in the direction of its antipodes. Bright pure colors are of the essence, not of beauty in general, but only of a special kind of beauty, the visionary. Gothic churches and Greek temples, the statues of the thirteenth century after Christ and of the fifth century before Christ—all were brilliantly colored.

For the Greeks and the men of the Middle Ages, this art of the merry-go-round and the waxwork show was evidently transporting. To us it seems deplorable. We prefer our Praxiteleses plain, our marble and our limestone *au naturel*. Why should our modern taste be so different, in this respect, from that of our ancestors? The reason, I presume, is that we have become too familiar with bright, pure pigments to be greatly moved by them. We admire them, of course, when we see them in some grand or subtle composition; but in themselves and as such, they leave us untransported.

Sentimental lovers of the past complain of the drabness of our age and contrast it unfavorably with the gay brilliance of earlier times. In actual fact, of course, there is a far greater profusion of color in the modern than in the ancient world. Lapis lazuli and Tyrian purple were costly rarities; the rich velvets and brocades of princely wardrobes, the woven or painted hangings of medieval and early modern houses were reserved for a privileged minority.

Even the great ones of the earth possessed very few of these vision-inducing treasures. As late as the seventeenth century, monarchs owned so little furniture that they had to travel from palace to palace with wagonloads of plates and bedspreads, of carpets and tapestries. For the great mass of the people there were only homespun and a few vegetable dyes; and, for interior decoration, there were at best the earth colors, at worst (and in most cases) "the floor of plaster and the walls of dung."

At the antipodes of every mind lay the Other World of preternatural light and preternatural color, of ideal gems and visionary gold. But before every pair of eyes was only the dark squalor of the family hovel, the dust or mud of the village street, the dirty whites, the duns and goose-turd greens of ragged clothing. Hence a passionate, an almost desperate thirst for bright, pure colors; and hence the overpowering effect produced by such colors whenever, in church or at court, they were displayed. Today the chemical industry turns out paints, inks and dyes in endless variety and enormous quantities. In our modern world there is enough bright color to guarantee the production of billions of flags and comic strips, millions of stop signs and taillights, fire engines and Coca-Cola containers by the hundred thousand, carpets, wallpapers and non-representational art by the square mile.

Familiarity breeds indifference. We have seen too much pure, bright color at Woolworth's to find it intrinsically transporting. And here we may note that, by its amazing capacity to give us too much of the best things, modern technology has tended to devaluate the traditional vision-inducing materials. The illumination of a city, for example, was once a rare event, reserved for victories and national holidays, for the canonization of saints and the crowning of kings. Now it occurs nightly and celebrates the virtues of gin, cigarettes and toothpaste.

In London, fifty years ago, electric sky signs were a novelty and so rare that they shone out of the misty darkness "like captain jewels in the carcanet." Across the Thames, on the old Shot Tower, the gold and ruby letters were magically lovely—*une féerie*. Today the fairies are gone. Neon is everywhere and, being everywhere, has no effect upon us, except perhaps to make us pine nostalgically for primeval night.

Only in floodlighting do we recapture the unearthly significance which used, in the age of oil and wax, even in the age of gas and the carbon filament, to shine forth from practically any island of brightness in the boundless dark. Under the searchlights Notre Dame de Paris and the Roman Forum are visionary objects, having power to transport the beholder's mind toward the Other World.*

Modern technology has had the same devaluating effect on glass and polished metal as it has had on fairy lamps and pure, bright colors. By John of Patmos and his contemporaries walls of glass were conceivable only in the New Jerusalem. Today they are a feature of every up-to-date office, building and bungalow. And this glut of glass has been paralleled by a glut of chrome and nickel,

* See Appendix III, page 157.

of stainless steel and aluminum and a host of alloys old and new. Metal surfaces wink at us in the bathroom, shine from the kitchen sink, go glittering across country in cars and streamliners.

Those rich convex reflections, which so fascinated Rembrandt that he never tired of rendering them in paint, are now the commonplaces of home and street and factory. The fine point of seldom pleasure has been blunted. What was once a needle of visionary delight has now become a piece of disregarded linoleum.

I have spoken so far only of vision-inducing materials and their psychological devaluation by modern technology. It is time now to consider the purely artistic devices, by means of which vision-inducing works have been created.

Light and color tend to take on a preternatural quality when seen in the midst of environing darkness. Fra Angelico's "Crucifixion" at the Louvre has a black background. So have the frescoes of the Passion painted by Andrea del Castagno for the nuns of Sant' Appollonia at Florence. Hence the visionary intensity, the strange transporting power of these extraordinary works. In an entirely different artistic and psychological context the same device was often used by Goya in his etchings. Those flying men, that horse on the tightrope, the huge and ghastly incarnation of Fear—all of them stand out, as though floodlit, against a background of impenetrable night.

With the development of chiaroscuro, in the sixteenth and seventeenth centuries, night came out of the background and installed itself within the picture, which became the scene of a kind of Manichean struggle between Light and Darkness. At the time they were painted these works must have possessed a real transporting power. To us, who have seen altogether too much of this kind of thing, most of them seem merely theatrical. But a few still retain their magic. There is Caravaggio's "Entombment," for example; there are a dozen magical paintings by Georges de Latour;* there are all those visionary Rembrandts where the lights have the intensity and significance of light at the mind's antipodes, where the darks are full of rich potentialities waiting their turn to become actual, to make themselves glowingly present to our consciousness.

In most cases the ostensible subject matter of Rembrandt's pictures is taken from real life or the Bible—a boy at his lessons or Bathsheba bathing; a woman wading in a pond or Christ before His judges. Occasionally, how-

* See Appendix IV, page 173.

ever, these messages from the Other World are transmitted by means of a subject drawn, not from real life or history, but from the realm of archetypal symbols. There hangs in the Louvre a "Méditation du Philosophe," whose symbolical subject matter is nothing more nor less than the human mind, with its teeming darknesses, its moments of intellectual and visionary illumination, its mysterious stairways winding downward and upward into the unknown. The meditating philosopher sits there in his island of inner illumination; and at the opposite end of the symbolic chamber, in another, rosier island, an old woman crouches before the hearth. The firelight touches and transfigures her face, and we see, concretely illustrated, the impossible paradox and supreme truth— that perception is (or at least can be, ought to be) the same as Revelation, that Reality shines out of every appearance, that the One is totally, infinitely present in all particulars.

Along with the preternatural lights and colors, the gems and the ever-changing patterns, visitors to the mind's antipodes discover a world of sublimely beautiful landscapes, of living architecture and of heroic figures. The transporting power of many works of art is attributable to the fact that their creators have painted scenes, persons

ALDOUS HUXLEY

and objects which remind the beholder of what, consciously or unconsciously, he knows about the Other World at the back of his mind.

Let us begin with the human or, rather, the more than human inhabitants of these far-off regions. Blake called them the Cherubim. And in effect that is what, no doubt, they are—the psychological originals of those beings who, in the theology of every religion, serve as intermediaries between man and the Clear Light. The more than human personages of visionary experience never "do anything." (Similarly the blessed never "do anything" in heaven.) They are content merely to exist.

Under many names and attired in an endless variety of costumes, these heroic figures of man's visionary experience have appeared in the religious art of every culture. Sometimes they are shown at rest, sometimes in historical or mythological action. But action, as we have seen, does not come naturally to the inhabitants of the mind's antipodes. To be busy is the law of *our* being. The law of *theirs* is to do nothing. When we force these serene strangers to play a part in one of our all too human dramas, we are being false to visionary truth. That is why the most transporting (though not necessarily the most beautiful) representation of "the Cherubim" are those

which show them as they are in their native habitat—
doing nothing in particular.

And that accounts for the overwhelming, the more
than merely aesthetic impression made upon the beholder
by the great static masterpieces of religious art. The
sculptured figures of Egyptian gods and god-kings, the
Madonnas and Pantocrators of the Byzantine mosaics,
the bodhisattvas, and lohans of China, the seated Buddhas
of Khmer, the steles and statues of Copán, the wooden
idols of tropical Africa—these have one characteristic
in common: a profound stillness. And it is precisely this
which gives them their numinous quality, their power to
transport the beholder out of the old world of his everyday experience, far away, toward the visionary antipodes
of the human psyche.

There is, of course, nothing intrinsically excellent about
static art. Static or dynamic, a bad piece of work is
always a bad piece of work. All I mean to imply is that,
other things being equal, a heroic figure at rest has a
greater transporting power than one which is shown in
action.

The Cherubim live in Paradise and the New Jerusalem
—in other words, among prodigious buildings set in
rich, bright gardens with distant prospects of plain and

mountain, of rivers and the sea. This is a matter of immediate experience, a psychological fact which has been recorded in folklore and the religious literature of every age and country. It has not, however, been recorded in pictorial art.

Reviewing the succession of human cultures, we find that landscape painting is either non-existent, or rudimentary, or of very recent development. In Europe a full-blown art of landscape painting has existed for only four or five centuries, in China for not more than a thousand years, in India, for all practical purposes, never.

This is a curious fact that demands an explanation. Why should landscapes have found their way into the visionary literature of a given epoch and a given culture, but not into the painting? Posed in this way, the question provides its own best answer. People may be content with the merely verbal expression of this aspect of their visionary experience and feel no need for its translation into pictorial terms.

That this often happens in the case of individuals is certain. Blake, for example, saw visionary landscapes "articulated beyond all that the mortal and perishing nature can produce" and "infinitely more perfect and minutely organized than anything seen by the mortal

eye." Here is the description of such a visionary landscape, which Blake gave at one of Mrs. Aders' evening
parties: "The other evening, taking a walk, I came to a
meadow and at the further corner of it I saw a fold of
lambs. Coming nearer, the ground blushed with flowers,
and the wattled cote and its woolly tenants were of an
exquisite pastoral beauty. But I looked again, and it
proved to be no living flock, but beautiful sculpture."

Rendered in pigments, this vision would look, I suppose, like some impossibly beautiful blending of one of
Constable's freshest oil sketches with an animal painting
in the magically realistic style of Zurbarán's haloed lamb
now in the San Diego Museum. But Blake never produced
anything remotely resembling such a picture. He was
content to talk and write about his landscape visions, and
to concentrate in his painting upon "the Cherubim."

What is true of an individual artist may be true of a
whole school. There are plenty of things which men
experience, but do not choose to express; or they may try
to express what they have experienced, but in only one
of their arts. In yet other cases they will express themselves in ways having no immediately recognizable affinity
to the original experience. In this last context Dr. A. K.
Coomaraswamy has some interesting things to say about

the mystical art of the Far East—the art where "denotation and connotation cannot be divided" and "no distinction is felt between what a thing 'is' and what it 'signifies.' "

The supreme example of such mystical art is the Zen-inspired landscape painting which arose in China during the Sung period and came to new birth in Japan four centuries later. India and the Near East have no mystical landscape painting; but they have its equivalents— "Vaisnava painting, poetry and music in India, where the theme is sexual love; and Sufi poetry and music in Persia, devoted to praises of intoxication."*

"Bed," as the Italian proverb succinctly puts it, "is the poor man's opera." Analogously, sex is the Hindu's Sung; wine, the Persian's Impressionism. The reason being, of course, that the experiences of sexual union and intoxication partake of that essential otherness characteristic of all vision, including that of landscapes.

If, at any time, men have found satisfaction in a certain kind of activity, it is to be presumed that, at periods when this satisfying activity was not manifested, there must have been some kind of equivalent for it. In

* A. K. Coomaraswamy, *The Transformation of Nature in Art*, p. 40.

the Middle Ages, for example, men were preoccupied in an obsessive, an almost maniacal way with words and symbols. Everything in nature was instantly recognized as the concrete illustration of some notion formulated in one of the books or legends currently regarded as sacred.

And yet, at other periods of history men have found a deep satisfaction in recognizing the autonomous otherness of nature, including many aspects of human nature. The experience of this otherness was expressed in terms of art, religion or science. What were the medieval equivalents of Constable and ecology, of bird watching and Eleusis, of microscopy and the rites of Dionysus and the Japanese Haiku? They were to be found, I suspect, in Saturnalian orgies at one end of the scale and in mystical experience at the other. Shrovetides, May Days, Carnivals—these permitted a direct experience of the animal otherness underlying personal and social identity. Infused contemplation revealed the yet otherer otherness of the divine Not-Self. And somewhere between the two extremes were the experiences of the visionaries and the vision-inducing arts, by means of which it was sought to recapture and re-create those experiences—the art of the jeweler, of the maker of stained glass, of the weaver of tapestries, of the painter, poet and musician.

In spite of a natural history that was nothing but a set of drearily moralistic symbols, in the teeth of a theology which, instead of regarding words as the signs of things, treated things and events as the signs of Biblical or Aristotelian words, our ancestors remained relatively sane. And they achieved this feat by periodically escaping from the stifling prison of their bumptiously rationalistic philosophy, their anthropomorphic, authoritarian and non-experimental science, their all too articulate religion, into non-verbal, other than human worlds inhabited by their instincts, by the visionary fauna of their mind's antipodes and, beyond and yet within all the rest, by the indwelling Spirit.

From this wide-ranging but necessary digression, let us return to the particular case from which we set out. Landscapes, as we have seen, are a regular feature of the visionary experience. Descriptions of visionary landscapes occur in the ancient literature of folklore and religion; but paintings of landscapes do not make their appearance until comparatively recent times. To what has been said, by way of explanation about psychological equivalents, I will add a few brief notes on the nature of landscape painting as a vision-inducing art.

Let us begin by asking a question. What landscapes—or, more generally, what representations of natural objects—are most transporting, most intrinsically vision inducing? In the light of my own experience and of what I have heard other people say about their reactions to works of art, I will risk an answer. Other things being equal (for nothing can make up for lack of talent), the most transporting landscapes are, first, those which represent natural objects a very long way off, and, second, those which represent them at close range.

Distance lends enchantment to the view; but so does propinquity. A Sung painting of faraway mountains, clouds and torrents is transporting; but so are the close-ups of tropical leaves in the Douanier Rousseau's jungles. When I look at the Sung landscape, I am reminded (or one of my not-I's is reminded) of the crags, the boundless expanses of plain, the luminous skies and seas of the mind's antipodes. And those disappearances into mist and cloud, those sudden emergences of some strange, intensely definite form, a weathered rock, for example, an ancient pine tree twisted by years of struggle with the wind—these too, are transporting. For they remind me, consciously or unconsciously, of the Other World's essential alienness and unaccountability.

It is the same with the close-up. I look at those leaves with their architecture of veins, their stripes and mottlings, I peer into the depths of interlacing greenery, and something in me is reminded of those living patterns, so characteristic of the visionary world, of those endless births and proliferations of geometrical forms that turn into objects, of things that are forever being transmuted into other things.

This painted close-up of a jungle is what, in one of its aspects, the Other World is like, and so it transports me, it makes me see with eyes that transfigure a work of art into something else, something beyond art.

I remember—very vividly. though it took place many years ago—a conversation with Roger Fry. We were talking about Monet's "Water Lilies." They had no right, Roger kept insisting, to be so shockingly unorganized, so totally without a proper compositional skeleton. They were all wrong, artistically speaking. And yet, he had to admit, and yet. . . . And yet, as I should now say, they were transporting. An artist of astounding virtuosity had chosen to paint a close-up of natural objects seen in their own context and without reference to merely human notions of what's what, or what ought to be what. Man, we like to say, is the measure of all things. For Monet, on this occasion, water lilies were the measure of water lilies; and so he painted them.

The same non-human point of view must be adopted by any artist who tries to render the distant scene. How tiny, in the Chinese painting, are the travelers who make their way along the valley! How frail the bamboo hut on the slope above them! And all the rest of the vast landscape is emptiness and silence. This revelation of the wilderness, living its own life according to the laws of its own being, transports the mind toward its antipodes; for primeval Nature bears a strange resemblance to that inner world where no account is taken of our personal wishes or even of the enduring concerns of man in general.

Only the middle distance and what may be called the remoter foreground are strictly human. When we look very near or very far, man either vanishes altogether or loses his primacy. The astronomer looks even further afield than the Sung painter and sees even less of human life. At the other end of the scale the physicist, the chemist, the physiologist pursue the close-up—the cellular close-up, the molecular, the atomic and sub-atomic. Of that which, at twenty feet, even at arm's length, looked and sounded like a human being no trace remains.

Something analogous happens to the myopic artist and the happy lover. In the nuptial embrace personality is melted down; the individual (it is the recurrent theme of Lawrence's poems and novels) ceases to be himself and becomes a part of the vast impersonal universe.

And so it is with the artist who chooses to use his eyes at the near point. In his work humanity loses its importance, even disappears completely. Instead of men and women playing their fantastic tricks before high heaven, we are asked to consider the lilies, to meditate on the unearthly beauty of "mere things," when isolated from their utilitarian context and rendered as they are, in and for themselves. Alternatively (or, at an earlier stage of artistic development, exclusively) the non-human world of the near point is rendered in patterns. These patterns are abstracted for the most part from leaves and flowers —the rose, the lotus, the acanthus, palm, papyrus—and are elaborated, with recurrences and variations, into something transportingly reminiscent of the living geometries of the Other World.

Freer and more realistic treatments of Nature at the near point make their appearance at a relatively recent date—but far earlier than those treatments of the distant scene, to which alone (and mistakenly) we give the name of landscape painting. Rome, for example, had its close-up landscapes. The fresco of a garden, which once adorned a room in Livia's villa, is a magnificent example of this form of art.

For theological reasons, Islam had to be content, for the most part, with "arabesques"—luxuriant and (as in visions) continually varying patterns, based upon natural objects seen at the near point. But even in Islam the genuine close-up landscape was not unknown. Nothing can exceed in beauty and in vision-inducing power the mosaics of gardens and buildings in the great Omayyad mosque at Damascus.

In medieval Europe, despite the prevailing mania for turning every datum into a concept, every immediate experience into a mere symbol of something in a book, realistic close-ups of foliage and flowers were fairly common. We find them carved on the capitals of Gothic pillars, as in the Chapter House of Southwell Cathedral. We find them in paintings of the chase—paintings whose subject was that ever-present fact of medieval life, the forest, seen as the hunter or the strayed traveler sees it, in all its bewildering intricacy of leafy detail.

131

The frescoes in the papal palace at Avignon are almost the sole survivors of what, even in the time of Chaucer, was a widely practiced form of secular art. A century later this art of the forest close-up came to its self-conscious perfection in such magnificent and magical works as Pisanello's "St. Hubert" and Paolo Uccello's "Hunt in a Wood," now in the Ashmolean Museum at Oxford. Closely related to the wall paintings of forest close-ups were the tapestries, with which the rich men of northern Europe adorned their houses. The best of these are vision-inducing works of the highest order. In their own way they are as heavenly, as powerfully reminiscent of what goes on at the mind's antipodes, as are the great master-pieces of landscape painting at the farthest point—Sung mountains in their enormous solitude, Ming rivers inter-minably lovely, the blue sub-Alpine world of Titian's dis-tances, the England of Constable; the Italies of Turner and Corot; the Provences of Cézanne and Van Gogh; the Île de France of Sisley and the Île de France of Vuillard.

Vuillard, incidentally, was a supreme master both of the transporting close-up and of the transporting distant view. His bourgeois interiors are masterpieces of vision-inducing art, compared with which the works of such conscious and so to say professional visionaries as Blake and Odilon Redon seem feeble in the extreme. In Vuillard's interior every detail however trivial, however hideous even—the pattern of the late Victorian wallpaper, the *art nouveau* bibelot, the Brussels carpet—is seen and rendered as a living jewel; and all these jewels are harmoniously combined into a whole which is a jewel of a yet higher order of visionary intensity. And when the upper middle-class inhabitants of Vuillard's New Jerusalem go for a walk, they find themselves not, as they had supposed, in the department of Seine-et-Oise, but in the Garden of Eden, in an Other World which is yet essentially the same as this world, but transfigured and therefore transporting.*

I have spoken so far only of the blissful visionary experience and of its interpretation in terms of theology, its translation into art. But visionary experience is not always blissful. It is sometimes terrible. There is hell as well as heaven.

Like heaven, the visionary hell has its preternatural light and its preternatural significance. But the significance is intrinsically appalling and the light is "the smoky light" of the *Tibetan Book of the Dead*, the "darkness

* See Appendix V, page 176.

visible" of Milton. In the *Journal d'une Schizophrène,** the autobiographical record of a young girl's passage through madness, the world of the schizophrenic is called *le Pays d'Éclairement*—"the country of lit-upness." It is a name which a mystic might have used to denote his heaven. But for poor Renée, the schizophrenic, the illumination is infernal—an intense electric glare without a shadow, ubiquitous and implacable. Everything that, for healthy visionaries, is a source of bliss brings to Renée only fear and a nightmarish sense of unreality. The summer sunshine is malignant; the gleam of polished surfaces is suggestive not of gems, but of machinery and enameled tin; the intensity of existence which animates every object, when seen at close range and out of its utilitarian context, is felt as a menace.

And then there is the horror of infinity. For the healthy visionary, the perception of the infinite in a finite particular is a revelation of divine immanence; for Renée, it was a revelation of what she calls "the System," the vast cosmic mechanism which exists only to grind out guilt and punishment, solitude and unreality.**

Sanity is a matter of degree, and there are plenty of

* *Journal d'une Schizophrène,* by M. A. Sechehaye. Paris. 1950.
** See Appendix VI, page 180.

visionaries, who see the world as Renée saw it, but contrive, none the less, to live outside the asylum. For them, as for the positive visionary, the universe is transfigured —but for the worse. Everything in it, from the stars in the sky to the dust under their feet, is unspeakably sinister or disgusting; every event is charged with a hateful significance; every object manifests the presence of an Indwelling Horror, infinite, all-powerful, eternal.

This negatively transfigured world has found its way, from time to time, into literature and the arts. It writhed and threatened in Van Gogh's later landscapes; it was the setting and the theme of all Kafka's stories; it was Géricault's spiritual home;* it was inhabited by Goya during the long years of his deafness and solitude; it was glimpsed by Browning when he wrote *Childe Roland;* it had its place, over against the theophanies, in the novels of Charles Williams.

The negative visionary experience is often accompanied by bodily sensations of a very special and characteristic kind. Blissful visions are generally associated with a sense of separation from the body, a feeling of deindividualization. (It is, no doubt, this feeling of deindividualization which makes it possible for the Indians who practice the

* See Appendix VII, page 182.

peyote cult to use the drug not merely as a short cut to the visionary world, but also as an instrument for creating a loving solidarity within the participating group.) When the visionary experience is terrible and the world is transfigured for the worse, individualization is intensified and the negative visionary finds himself associated with a body that seems to grow progressively more dense, more tightly packed, until he finds himself at last reduced to being the agonized consciousness of an inspissated lump of matter, no bigger than a stone that can be held between the hands.

It is worth remarking, that many of the punishments described in the various accounts of hell are punishments of pressure and constriction. Dante's sinners are buried in mud, shut up in the trunks of trees, frozen solid in blocks of ice, crushed beneath stones. The *Inferno* is psychologically true. Many of its pains are experienced by schizophrenics, and by those who have taken mescalin or lysergic acid under unfavorable conditions.*

What is the nature of these unfavorable conditions? How and why is heaven turned into hell? In certain cases the negative visionary experience is the result of predominantly physical causes. Mescalin tends, after

* See Appendix VIII, page 184.

ingestion, to accumulate in the liver. If the liver is diseased, the associated mind may find itself in hell. But what is more important for our present purposes is the fact that negative visionary experience may be induced by purely psychological means. Fear and anger bar the way to the heavenly Other World and plunge the mescalin taker into hell.

And what is true of the mescalin taker is also true of the person who sees visions spontaneously or under hypnosis. Upon this psychological foundation has been reared the theological doctrine of saving faith—a doctrine to be met with in all the great religious traditions of the world. Eschatologists have always found it difficult to reconcile their rationality and their morality with the brute facts of psychological experience. As rationalists and moralists, they feel that good behavior should be rewarded and that the virtuous deserve to go to heaven. But as psychologists they know that virtue is not the sole or sufficient condition of blissful visionary experience. They know that works alone are powerless and that it is faith, or loving confidence, which guarantees that visionary experience shall be blissful.

Negative emotions—the fear which is the absence of confidence, the hatred, anger or malice which exclude love—are the guarantee that visionary experience, if and when it comes, shall be appalling. The Pharisee is a virtuous man; but his virtue is of the kind which is compatible with negative emotion. His visionary experiences are therefore likely to be infernal rather than blissful.

The nature of the mind is such that the sinner who repents and makes an act of faith in a higher power is more likely to have a blissful visionary experience than is the self-satisfied pillar of society with his righteous indignations, his anxiety about possessions and pretensions, his ingrained habits of blaming, despising and condemning. Hence the enormous importance attached, in all the great religious traditions, to the state of mind at the moment of death.

Visionary experience is not the same as mystical experience. Mystical experience is beyond the realm of opposites. Visionary experience is still within that realm. Heaven entails hell, and "going to heaven" is no more liberation than is the descent into horror. Heaven is merely a vantage point, from which the divine Ground can be more clearly seen than on the level of ordinary individualized existence.

If consciousness survives bodily death, it survives,

presumably, on every mental level—on the level of mystical experience, on the level of blissful visionary experience, on the level of infernal visionary experience, and on the level of everyday individual existence. In life, as we know by experience and observation, even the blissful visionary experience tends to change its sign if it persists too long.

Many schizophrenics have their times of heavenly happiness; but the fact that (unlike the mescalin taker) they do not know when, if ever, they will be permitted to return to the reassuring banality of everyday experience causes even heaven to seem appalling. But for those who, for whatever reason, are appalled, heaven turns into hell, bliss into horror, the Clear Light into the hateful glare of the land of lit-upness.

Something of the same kind may happen in the posthumous state. After having had a glimpse of the unbearable splendor of ultimate Reality, and after having shuttled back and forth between heaven and hell, most souls find it possible to retreat into that more reassuring region of the mind, where they can use their own and other people's wishes, memories and fancies to construct a world very like that in which they lived on earth.

Of those who die an infinitesimal minority are capable

of immediate union with the divine Ground, a few are
capable of supporting the visionary bliss of heaven, a few
find themselves in the visionary horrors of hell and are
unable to escape; the great majority end up in the kind
of world described by Swedenborg and the mediums.
From this world it is doubtless possible to pass, when the
necessary conditions have been fulfilled, to worlds of
visionary bliss or the final enlightenment.

My own guess is that modern spiritualism and ancient
tradition are both correct. There *is* a posthumous state
of the kind described in Sir Oliver Lodge's book *Raymond;* but there is also a heaven of blissful visionary
experience; there is also a hell of the same kind of
appalling visionary experience as is suffered here by
schizophrenics and some of those who take mescalin; and
there is also an experience, beyond time, of union with
the divine Ground.

APPENDICES

I.

Two other, less effective aids to vision-ary experience deserve mention—carbon dioxide and the stroboscopic lamp. A mixture (completely non-toxic) of seven parts of oxygen and three of carbon dioxide produces, in those who inhale it, certain physical and psychological changes, which have been exhaustively described by Meduna. Among these changes the most important, in our present context, is a marked enhance-ment of the ability to "see things," when the eyes are closed. In some cases only swirls of patterned color are seen. In others there may be vivid recalls of past experi-ences. (Hence the value of CO_2 as a therapeutic agent.) In yet other cases carbon dioxide transports the subject to the Other World at the antipodes of his everyday con-sciousness, and he enjoys very briefly visionary experi-ences entirely unconnected with his own personal history or with the problems of the human race in general.

In the light of these facts it becomes easy to understand

the rationale of yogic breathing exercises. Practiced systematically, these exercises result, after a time, in prolonged suspensions of breath. Long suspensions of breath lead to a high concentration of carbon dioxide in the lungs and blood, and this increase in the concentration of CO_2 lowers the efficiency of the brain as a reducing valve and permits the entry into consciousness of experiences, visionary or mystical, from "out there."

Prolonged and continuous shouting or singing may produce similar, but less strongly marked, results. Unless they are highly trained, singers tend to breathe out more than they breathe in. Consequently the concentration of carbon dioxide in the alveolar air and the blood is increased and, the efficiency of the cerebral reducing valve being lowered, visionary experience becomes possible. Hence the interminable "vain repetitions" of magic and religion. The chanting of the *curandero*, the medicine man, the shaman; the endless psalm singing and sutra intoning of Christian and Buddhist monks; the shouting and howling, hour after hour, of revivalists—under all the diversities of theological belief and aesthetic convention, the psychochemico-physiological intention remains constant. To increase the concentration of CO_2 in the lungs and blood and so to lower the efficiency of the cerebral reducing valve, until it will admit biologically useless material from Mind-at-Large—this, though the shouters, singers and mutterers did not know it, has been at all times the real purpose and point of magic spells, of mantrams, litanies, psalms and sutras. "The heart," said Pascal, "has its reasons." Still more cogent and much harder to unravel are the reasons of the lungs, the blood and the enzymes, of neurons and synapses. The way to the superconscious is through the subconscious, and the way, or at least one of the ways, to the subconscious is through the chemistry of individual cells.

With the stroboscopic lamp we descend from chemistry to the still more elementary realm of physics. Its rhythmically flashing light seems to act directly, through the optic nerves, on the electrical manifestations of the brain's activity. (For this reason there is always a slight danger involved in the use of the stroboscopic lamp. Some persons suffer from *petit mal* without being made aware of the fact by any clear-cut and unmistakable symptoms. Exposed to a stroboscopic lamp, such persons may go into a full-blown epileptic fit. The risk is not very great; but it must always be recognized. One case in eighty may turn out badly.)

To sit, with eyes closed, in front of a stroboscopic

lamp is a very curious and fascinating experience. No sooner is the lamp turned on than the most brilliantly colored patterns make themselves visible. These patterns are not static, but change incessantly. Their prevailing color is a function of the stroboscope's rate of discharge. When the lamp is flashing at any speed between ten to fourteen or fifteen times a second, the patterns are prevailingly orange and red. Green and blue make their appearance when the rate exceeds fifteen flashes a second. After eighteen or nineteen, the patterns become white and gray. Precisely why we should see such patterns under the stroboscope is not known. The most obvious explanation would be in terms of the interference of two or more rhythms—the rhythm of the lamp and the various rhythms of the brain's electrical activity. Such interferences may be translated by the visual center and optic nerves into something of which the mind becomes conscious as a colored, moving pattern. Far more difficult to explain is the fact, independently observed by several experimenters, that the stroboscope tends to enrich and intensify the visions induced by mescalin or lysergic acid. Here, for example, is a case communicated to me by a medical friend. He had taken lysergic acid and was seeing, with his eyes shut, only colored, moving patterns. Then he sat down in front of a stroboscope. The lamp was turned on and, immediately, abstract geometry was transformed into what my friend described as "Japanese landscapes" of surpassing beauty. But how on earth can the interference of two rhythms produce an arrangement of electrical impulses interpretable as a living, self-modulating Japanese landscape unlike anything the subject has ever seen, suffused with preternatural light and color and charged with preternatural significance?

This mystery is merely a particular case of a larger, more comprehensive mystery—the nature of the relations between visionary experience and events on the cellular, chemical and electrical levels. By touching certain areas of the brain with a very fine electrode, Penfield has been able to induce the recall of a long chain of memories relating to some past experience. This recall is not merely accurate in every perceptual detail; it is also accompanied by all the emotions which were aroused by the events when they originally occurred. The patient, who is under a local anesthetic, finds himself simultaneously in two times and places—in the operating room, now, and in his childhood home, hundreds of miles away and thousands of days in the past. Is there, one wonders, some area in the brain from which the probing electrode could elicit Blake's Cherubim, or Weir Mitchell's self-transformating Gothic tower encrusted with living gems, or my friend's unspeakably lovely Japanese landscapes? And if, as I myself believe, visionary experiences enter our consciousness from somewhere "out there" in the infinity of Mind-at-Large, what sort of an *ad hoc* neurological pattern is created for them by the receiving and transmitting brain? And what happens to this *ad hoc* pattern when the vision is over? Why do all visionaries insist on the impossibility of recalling, in anything even faintly resembling its original form and intensity, their transfiguring experiences? How many questions—and, as yet, how few answers!

I I.

In the Western world, visionaries and mystics are a good deal less common than they used to be. There are two principal reasons for this state of affairs—a philosophical reason and a chemical reason. In the currently fashionable picture of the universe there is no place for valid transcendental experience. Consequently those who have had what they regard as valid transcendental experiences are looked upon with suspicion as being either lunatics or swindlers. To be a mystic or a visionary is no longer creditable.

But it is not only our mental climate that is unfavorable to the visionary and the mystic; it is also our chemical environment—an environment profoundly different from that in which our forefathers passed their lives.

The brain is chemically controlled, and experience has shown that it can be made permeable to the (biologically speaking) superfluous aspects of Mind-at-Large by modifying the (biologically speaking) normal chemistry of the body.

For almost half of every year our ancestors ate no fruit, no green vegetables and (since it was impossible for them to feed more than a few oxen, cows, swine and poultry during the winter months) very little butter or fresh meat, and very few eggs. By the beginning of each successive spring, most of them were suffering, mildly or acutely, from scurvy, due to lack of Vitamin C, and pellagra, caused by a shortage in their diet of the B complex. The distressing physical symptoms of these diseases are associated with no less distressing psychological symptoms.* The nervous system is more vulnerable than the other tissues of the body; consequently vitamin deficiencies tend to affect the state of mind before they affect, at least in any very obvious way, the skin, bones, mucous membranes, muscles and viscera. The first result of an inadequate diet is a lowering of the efficiency of the brain as an instrument for biological survival. The undernourished person tends to be afflicted by anxiety, depression, hypochondria and feelings of anxiety. He is also liable to see visions; for when the cerebral reducing valve has its efficiency lowered, much (biologically speaking) useless

* See *The Biology of Human Starvation* by A. Keys (University of Minnesota Press, 1950); also the recent (1955) reports of the work on the role of vitamin deficiencies in mental disease, carried out by Dr. George Watson and his associates in Southern California.

material flows into consciousness from "out there," in Mind-at-Large.

Much of what the earlier visionaries experienced was terrifying. To use the language of Christian theology, the Devil revealed himself in their visions and ecstasies a good deal more frequently than did God. In an age when vitamins were deficient and a belief in Satan universal, this was not surprising. The mental distress, associated with even mild cases of pellagra and scurvy, was deepened by fears of damnation and a conviction that the powers of evil were omnipresent. This distress was apt to tinge with its own dark coloring the visionary material, admitted to consciousness through a cerebral valve whose efficiency had been impaired by underfeeding. But in spite of their preoccupations with eternal punishment and in spite of their deficiency disease, spiritually minded ascetics often saw heaven and might even be aware, occasionally, of that divinely impartial One in which the polar opposites are reconciled. For a glimpse of beatitude, for a foretaste of unitive knowledge, no price seemed too high. Mortification of the body may produce a host of undesirable mental symptoms; but it may also open a door into a transcendental world of Being, Knowledge and Bliss. That is why, in spite of its obvious disadvantages, almost all aspirants to the spiritual life have, in the past, undertaken regular courses of bodily mortification.

So far as vitamins were concerned, every medieval winter was a long involuntary fast, and this involuntary fast was followed, during Lent, by forty days of voluntary abstinence. Holy Week found the faithful marvelously well prepared, so far as their body chemistry was concerned, for its tremendous incitements to grief and joy, for seasonable remorse of conscious and a self-transcending identification with the risen Christ. At this season of the highest religious excitement and the lowest vitamin intake, ecstasies and visions were almost a commonplace. It was only to be expected.

For cloistered contemplatives, there were several Lents in every year. And even between fasts their diet was meager in the extreme. Hence those agonies of depression and scrupulosity described by so many spiritual writers; hence their frightful temptations to despair and self-slaughter. But hence too those "gratuitous graces" in the form of heavenly visions and locutions, of prophetic insights, of telepathic "discernments of spirits." And hence, finally, their "infused contemplation," their "obscure knowledge" of the One in all.

Fasting was not the only form of physical mortification

resorted to by the earlier aspirants to spirituality. Most of them regularly used upon themselves the whip of knotted leather or even of iron wire. These beatings were the equivalent of fairly extensive surgery without anesthetics, and their effects on the body chemistry of the penitent were considerable. Large quantities of histamine and adrenalin were released while the whip was actually being plied; and when the resulting wounds began to fester (as wounds practically always did before the age of soap), various toxic substances, produced by the decomposition of protein, found their way into the blood stream. But histamine produces shock, and shock affects the mind no less profoundly than the body. Moreover, large quantities of adrenalin may cause hallucinations, and some of the products of its decomposition are known to induce symptoms resembling those of schizophrenia. As for toxins from wounds—these upset the enzyme systems regulating the brain, and lower its efficiency as an instrument for getting on in a world where the biologically fittest survive. This may explain why the Curé d'Ars used to say that, in the days when he was free to flagellate himself without mercy, God would refuse him nothing. In other words, when remorse, self-loathing and the fear of hell release adrenalin, when self-inflicted surgery releases adrenalin and histamine, and when infected wounds release decomposed protein into the blood, the efficiency of the cerebral reducing valve is lowered and unfamiliar aspects of Mind-at-Large (including psi phenomena, visions and, if he is philosophically and ethically prepared for it, mystical experiences) will flow into the ascetic's consciousness.

Lent, as we have seen, followed a long period of involuntary fasting. Analogously, the effects of self-flagellation were supplemented, in earlier times, by much involuntary absorption of decomposed protein. Dentistry was non-existent, surgeons were executioners, and there were no safe antiseptics. Most people, therefore, must have lived out their lives with focal infections; and focal infections, though out of fashion as the cause of *all* the ills that flesh is heir to, can certainly lower the efficiency of the cerebral reducing valve.

And the moral of all this is—what? Exponents of a Nothing-But philosophy will answer that, since changes in body chemistry can create the conditions favorable to visionary and mystical experiences, visionary and mystical experiences cannot be what they claim to be, what, for those who have had them, they self-evidently are. But this, of course, is a *non sequitur*.

A similar conclusion will be reached by those whose philosophy is unduly "spiritual." God, they will insist, is a spirit and is to be worshiped in spirit. Therefore an experience which is chemically conditioned cannot be an experience of the divine. But, in one way or another, *all* our experiences are chemically conditioned, and if we imagine that some of them are purely "spiritual," purely "intellectual," purely "aesthetic," it is merely because we have never troubled to investigate the internal chemical environment at the moment of their occurrence. Furthermore, it is a matter of historical record that most contemplatives worked systematically to modify their body chemistry, with a view to creating the internal conditions favorable to spiritual insight. When they were not starving themselves into low blood sugar and a vitamin deficiency, or beating themselves into intoxication by histamine, adrenalin and decomposed protein, they were cultivating insomnia and praying for long periods in uncomfortable positions in order to create the psycho-physical symptoms of stress. In the intervals they sang interminable psalms, thus increasing the amount of carbon dioxide in the lungs and the blood stream, or, if they were Orientals, they did breathing exercises to accomplish the same purpose. Today we know how to lower the efficiency of the cerebral reducing valve by direct chemical action, and without the risk of inflicting serious damage on the psycho-physical organism. For an aspiring mystic to revert, in the present state of knowledge, to prolonged fasting and violent self-flagellation would be as senseless as it would be for an aspiring cook to behave like Charles Lamb's Chinaman, who burned down the house in order to roast a pig. Knowing as he does (or at least as he can know, if he so desires) what are the chemical conditions of transcendental experience, the aspiring mystic should turn for technical help to the specialists—in pharmacology, in biochemistry, in physiology and neurology, in psychology and psychiatry, and parapsychology. And on their part, of course, the specialists (if any of them aspire to be genuine men of science and complete human beings) should turn, out of their respective pigeonholes, to the artist, the sibyl, the visionary, the mystic—all those, in a word, who have had experience of the Other World and who know, in their different ways, what to do with that experience.

III.

Visionlike effects and vision-inducing devices have
played a greater part in popular entertainment than in
the fine arts. Fireworks, pageantry, theatrical spectacles—
those are essentially visionary arts. Unfortunately they
are also ephemeral arts, whose earlier masterpieces are
known to us only by report. Nothing remains of all
the Roman triumphs, the medieval tournaments, the
Jacobean masques, the long succession of state entries
and coronations, of royal marriages and solemn decapi-
tations, of canonizations and the funerals of Popes. The
best that can be hoped for such magnificences is that
they may "live in Settle's numbers one day more."

An interesting feature of these popular visionary arts
is their close dependence upon contemporary technology.
Fireworks, for example, were once no more than bonfires.
(And to this day, I may add, a good bonfire on a dark
night remains one of the most magical and transporting
of spectacles. Looking at it, one can understand the
mentality of the Mexican peasant, who sets out to burn an

acre of woodland in order to plant his maize, but is delighted when, by a happy accident, a square mile or two goes up in bright, apocalyptic flame.) True pyrotechny began (in Europe at least, if not in China) with the use of combustibles in sieges and naval battles. From war it passed, in due course, to entertainment. Imperial Rome had its firework displays, some of which, even in its decline, were elaborate in the extreme. Here is Claudian's description of the show put on by Manlius Theodorus in A.D. 399.

> Mobile ponderibus descendat pegma reductis
> inque chori speciem spargentes ardua flammas
> scaena rotet varios, et fingat Mulciber orbis
> per tabulas impune vagos pictaeque citato
> ludent igne trabes, et non permissa morari
> fida per innocuas errent incendia turres.

"Let the counterweights be removed," Mr. Platnauer translates with a straightforwardness of language that does less than justice to the syntactical extravagances of the original, "and let the mobile crane descend, lowering on to the lofty stage men who, wheeling chorus-wise, scatter flames. Let Vulcan forge balls of fire to roll innocuously across the boards. Let the flames appear to play about the sham beams of the scenery and a tame conflagration, never allowed to rest, wander among the untouched towers."

After the fall of Rome, pyrotechny became, once more, exclusively a military art. Its greatest triumph was the invention by Callinicus, about A.D. 650, of the famous Greek Fire—the secret weapon which enabled a dwindling Byzantine Empire to hold out for so long against its enemies.

With the Renaissance, fireworks re-entered the world of popular entertainment. With every advance in the science of chemistry, they became more and more brilliant. By the middle of the nineteenth century pyrotechny had reached a peak of technical perfection and was capable of transporting vast multitudes of spectators toward the visionary antipodes of minds which, consciously, were respectable Methodists, Puseyites, Utilitarians, disciples of Mill or Marx or Newman, or Bradlaugh, or Samuel Smiles. In the Piazza del Popolo, at Ranelagh and the Crystal Palace, on every Fourth and Fourteenth of July, the popular subconscious was reminded by the crimson glare of strontium, by copper blue and barium green and sodium yellow, of that Other World, down under, in the psychological equivalent of Australia.

Pageantry is a visionary art which has been used, from time immemorial, as a political instrument. The gorgeous fancy dress worn by kings, popes and their respective retainers, military and ecclesiastical, has a very practical purpose—to impress the lower classes with a lively sense of their masters' superhuman greatness. By means of fine clothes and solemn ceremonies *de facto* domination is transformed into a rule not merely *de jure,* but, positively, *de jure divino.* The crowns and tiaras, the assorted jewelry, the satins, silks and velvets, the gaudy uniforms and vestments, the crosses and medals, the sword hilts and the crosiers, the plumes in the cocked hats and their clerical equivalents, those huge feather fans which make every papal function look like a tableau from *Aïda*—all these are vision-inducing properties, designed to make all too human gentlemen and ladies look like heroes, demigoddesses and seraphs, and giving, in the process, a great deal of innocent pleasure to all concerned, actors and spectators alike.

In the course of the last two hundred years the technology of artificial lighting has made enormous progress, and this progress has contributed very greatly to the effectiveness of pageantry and the closely related art of theatrical spectacle. The first notable advance was made in the

eighteenth century, with the introduction of molded sper-
maceti candles in place of the older tallow dip and poured
wax taper. Next came the invention of Argand's tubular
wick, with an air supply on the inner as well as the outer
surface of the flame. Glass chimneys speedily followed,
and it became possible, for the first time in history, to
burn oil with a bright and completely smokeless light.
Coal gas was first employed as an illuminant in the early
years of the nineteenth century, and in 1825 Thomas
Drummond found a practical way of heating lime to
incandescence by means of an oxygen-hydrogen or oxy-
gen-coal gas flame. Meanwhile parabolic reflectors for
concentrating light into a narrow beam had come into
use. (The first English lighthouse equipped with such
a reflector was built in 1790.)

The influence on pageantry and theatrical spectacle
of these inventions was profound. In earlier times civic
and religious ceremonies could only take place during
the day (and days were as often cloudy as fine), or by
the light, after sunset, of smoky lamps and torches or the
feeble twinkling of candles. Argand and Drummond,
gas, limelight and, forty years later, electricity made it
possible to evoke, from the boundless chaos of night,
rich island universes, in which the glitter of metal and

gems, the sumptuous glow of velvets and brocades were intensified to the highest pitch of what may be called intrinsic significance. A recent example of ancient pageantry, raised by twentieth-century lighting to a higher magical power, was the coronation of Queen Elizabeth II. In the motion picture of the event, a ritual of transporting splendor was saved from the oblivion, which, up till now, has always been the fate of such solemnities, and preserved, blazing preternaturally under the floodlights, for the delight of a vast contemporary and future audience.

Two distinct and separate arts are practiced in the theater—the human art of the drama and the visionary, other-world art of spectacle. Elements of the two arts may be combined in a single evening's entertainment—the drama being interrupted (as so often happens in elaborate productions of Shakespeare) to permit the audience to enjoy a *tableau vivant* in which the actors either remain still or, if they move, move only in a non-dramatic way, ceremonially, processionally or in a formal dance. Our concern here is not with drama; it is with theatrical spectacle, which is simply pageantry without its political or religious overtones.

In the minor visionary arts of the costumier and the

designer of stage jewelry our ancestors were consummate masters. Nor, for all their dependence on unassisted muscle power, were they far behind us in the building and working of stage machinery, the contrivance of "special effects." In the masques of Elizabethan and early Stuart times, divine descents and irruptions of demons from the cellarage were a commonplace, so were apocalypses, so were the most amazing metamorphoses. Enormous sums of money were lavished on these spectacles. The Inns of Court, for example, put on a show for Charles I, which cost more than twenty thousand pounds —at a date when the purchasing power of the pound was six or seven times what it is today.

"Carpentry," said Ben Jonson sarcastically, "is the soul of masque." His contempt was motivated by resentment. Inigo Jones was paid as much for designing the scenery as was Ben for writing the libretto. The outraged laureate had evidently failed to grasp the fact that masque is a visionary art, and that visionary experience is beyond words (at any rate beyond all but the most Shakespearean words) and is to be evoked by direct, unmediated perceptions of things that remind the beholder of what is going on at the unexplored antipodes of his own personal consciousness. The soul of masque could

never, in the very nature of things, be a Jonsonian libretto; it *had* to be carpentry. But even carpentry could
not be the masque's whole soul. When it comes to us
from within, visionary experience is always preternaturally brilliant. But the early set designers possessed
no manageable illuminant brighter than a candle. At
close range a candle can create the most magical lights
and contrasting shadows. The visionary paintings of
Rembrandt and Georges de Latour are of things and
persons seen by candlelight. Unfortunately light obeys
the law of the inverse squares. At a safe distance from an
actor in inflammable fancy dress, candles are hopelessly
inadequate. At ten feet, for example, it would take one
hundred of the best wax tapers to produce an effective
illumination of one foot-candle. With such miserable
lighting only a fraction of the masque's visionary potentialities could be made actual. Indeed, its visionary potentialities were not fully realized until long after it had
ceased, in its original form, to exist. It was only in the
nineteenth century, when advancing technology had
equipped the theater with limelight and parabolic reflectors, that the masque came fully into its own. Victoria's
reign was the heroic age of the so-called Christmas pan-
tomime and the fantastic spectacle. "Ali Baba," "The

King of the Peacocks," "The Golden Branch," "The Island of Jewels"—their very names are magical. The soul of that theatrical magic was carpentry and dressmaking; its indwelling spirit, its *scintilla animæ*, was gas and lime-light and, after the eighties, electricity. For the first time in the history of the stage, beams of brightest incandescence transfigured the painted backdrops, the costumes, the glass and pinchbeck of jewelry, so that they became cap-able of transporting the spectators toward that Other World, which lies at the back of every mind, however perfect its adaptation to the exigencies of social life—even the social life of mid-Victorian England. Today we are in the fortunate position of being able to squander half a million horsepower on the nightly illumination of a metropolis. And yet, in spite of this devaluation of arti-ficial light, theatrical spectacle still retains its old com-pelling magic. Embodied in ballets, revues and musical comedies, the soul of masque goes marching along. Thousand-watt lamps and parabolic reflectors project beams of preternatural light, and preternatural light evokes, in everything it touches, preternatural color and preternatural significance. Even the silliest spectacle can be rather wonderful. It is a case of a New World having been called in to redress the balance of the Old—of visionary art making up for the deficiencies of all too human drama.

Athanasius Kircher's invention—if his, indeed, it was —was christened from the first *Lanterna Magica*. The name was everywhere adopted as perfectly appropriate to a machine, whose raw material was light, and whose finished product was a colored image emerging from the darkness. To make the original magic lantern show yet more magical, Kircher's successors devised a number of methods for imparting life and movement to the projected image. There were "chrometropic" slides in which two painted glass disks could be made to revolve in opposite directions, producing a crude but still effective imitation of those perpetually changing three-dimensional patterns, which have been seen by virtually everyone who has had a vision, whether spontaneous or induced by drugs, fasting or the stroboscopic lamp. Then there were those "dissolving views," which reminded the spectator of the metamorphoses going on incessantly at the antipodes of his everyday consciousness. To make one scene turn imperceptibly into another, two magic lanterns were used, projecting coincident images on the screen. Each lantern was fitted with a shutter, so arranged that the light of one could be progressively dimmed, while the light of the other (originally completely obscured) was progressively brightened. In this way the view projected by the first lantern was insensibly replaced by the view projected by the second—to the delight and astonishment of all beholders. Another device was the mobile magic lantern, projecting its image on a semi-transparent screen, on the further side of which sat the audience. When the lantern was wheeled close to the screen, the projected image was very small. As it was withdrawn, the image became progressively larger. An automatic focusing device kept the changing images sharp and unblurred at all distances. The word "phantasmagoria" was coined in 1802 by the inventors of this new kind of peepshow.

All these improvements in the technology of magic lanterns were contemporary with the poets and painters of the Romantic revival, and may perhaps have exercised a certain influence on their choice of subject matter and their methods of treating it. *Queen Mab* and *The Revolt of Islam,* for example, are full of dissolving views and phantasmagorias. Keat's descriptions of scenes and persons, of interiors and furniture and effects of light have the intense beamy quality of colored images on a white sheet in a darkened room. John Martin's representations of Satan and Belshazzar, of Hell and Babylon and the

Deluge are manifestly inspired by lantern slides and *tableaux vivants* dramatically illuminated by limelight.

The twentieth-century equivalent of the magic-lantern show is the colored movie. In the huge, expensive "spectaculars," the soul of masque goes marching along—with a vengeance sometimes, but sometimes also with taste and a real feeling for vision-inducing fantasy. Moreover, thanks to advancing technology, the colored documentary has proved itself, in skillful hands, a notable new form of popular visionary art. The immensely magnified cactus blossoms, into which, at the end of Disney's *The Living Desert,* the spectator finds himself sinking, come straight from the Other World. And then what transporting visions, in the best of the nature films, of foliage in the wind, of the textures of rock and sand, of the shadows and emerald lights in grass or among the reeds, of birds and insects and four-footed creatures going about their business in the underbrush or among the branches of forest trees! Here are the magical close-up landscapes which fascinated the makers of *mille-feuille* tapestries, the medieval painters of gardens and hunting scenes. Here are the enlarged and isolated details of living nature, out of which the artists of the Far East made some of the most beautiful of their paintings.

And then there is what may be called the Distorted Documentary—a new form of visionary art, admirably exemplified by Mr. Francis Thompson's film, *NY, NY*. In this very strange and beautiful picture we see the city of New York as it appears when photographed through multiplying prisms, or reflected in the backs of spoons, polished hub caps, spherical and parabolic mirrors. We still recognize houses, people, shop fronts, taxicabs, but recognize them as elements in one of those living geometries which are so characteristic of the visionary experience. The invention of this new cinematographic art seems to presage (thank heaven!) the supersession and early demise of non-representational painting. It used to be said by the non-representationalists that colored photography had reduced the old-fashioned portrait and the old-fashioned landscape to the rank of otiose absurdities. This, of course, is completely untrue. Colored photography merely records and preserves, in an easily reproducible form, the raw materials with which portraitists and landscape painters work. Used as Mr. Thompson has used it, colored cinematography does much more than merely record and preserve the raw materials of non-representational art; it actually turns out the finished product. Looking at *NY, NY*, I was amazed to see that virtually every pictorial device invented by the old masters of non-representational art and reproduced *ad nauseam* by the academicians and mannerists of the school, for the last forty years or more, makes its appearance, alive, glowing, intensely significant, in the sequences of Mr. Thompson's film.

Our ability to project a powerful beam of light has not only enabled us to create new forms of visionary art; it has also endowed one of the most ancient arts, the art of sculpture, with a new visionary quality which it did not previously possess. I have spoken in an earlier paragraph of the magical effects produced by the floodlighting of ancient monuments and natural objects. Analogous effects are seen when we turn the spotlights onto sculptured stone. Fuseli got the inspiration for some of his best and wildest pictorial ideas by studying the statues on Monte Cavallo by the light of the setting sun, or, better still, when illuminated by lightning flashes at midnight. Today we dispose of artificial sunsets and synthetic lightning. We can illuminate our statues from whatever angle we choose, and with practically any desired degree of intensity. Sculpture, in consequence, has revealed fresh meanings and unsuspected beauties. Visit the Louvre one night when the Greek and Egyptian antiquities are floodlit. You will meet with new gods, nymphs and Pharaohs; you will make the acquaintance, as one spotlight goes out and another, in a different quarter of space, is lit up, of a whole family of unfamiliar Victories of Samothrace.

The past is not something fixed and unalterable. Its facts are rediscovered by every succeeding generation, its values reassessed, its meanings redefined in the context of present tastes and preoccupations. Out of the same documents and monuments and works of art, every epoch invents its own Middle Ages, its private China, its patented and copyrighted Hellas. Today, thanks to recent advances in the technology of lighting, we can go one better than our predecessors. Not only have we reinterpreted the great works of sculpture bequeathed to us by the past, we have actually succeeded in altering the physical appearance of these works. Greek statues, as we see them illuminated by a light that never was on land or sea, and then photographed in a series of fragmentary close-ups from the oddest angles, bear almost no resemblance to the Greek statues seen by art critics and the general public in the dim galleries and decorous engravings of the past. The aim of the classical artist, in whatever period he may happen to live, is to impart order to the chaos of experience, to present a comprehensible, rational picture of reality, in which all the parts are clearly seen and coherently related, so that the beholder knows (or, to be more accurate, imagines that he knows) precisely what's what. To us, this ideal of rational orderliness makes no appeal. Consequently, when we are confronted by works of classical art, we use all the means in our power to make them look like something which they are not, and were never meant to be. From a work, whose whole point is its unity of conception, we select a single feature, focus our searchlights upon it and so force it, out of all context, upon the observer's consciousness. Where a contour seems to us too continuous, too obviously comprehensible, we break it up by alternating impenetrable shadows with patches of glaring brightness. When we photograph a sculptured figure or group, we use the camera to isolate a part, which we then exhibit in enigmatic independence from the whole. By such means we can de-classicize the severest classic. Subjected to the light treatment and photographed by an expert cameraman, a Pheidias becomes a piece of Gothic expressionism, a Praxiteles is turned into a fascinating *surrealist* object dredged up from the ooziest depths of the subconscious. This may be bad art history, but it is certainly enormous fun.

IV

Painter in ordinary, first to the Duke of his native Lorraine and later to the King of France, Georges de Latour was treated, during his lifetime, as the great artist he so manifestly was. With the accession of Louis XIV and the rise, the deliberate cultivation, of a new art of Versailles, aristocratic in subject matter and lucidly classical in style, the reputation of this once famous man suffered an eclipse so complete that, within a couple of generations, his very name had been forgotten, and his surviving paintings came to be attributed to the Le Nains, to Honthorst, to Zurbarán, to Murillo, even to Velázquez. The rediscovery of Latour began in 1915 and was virtually complete by 1934, when the Louvre organized a notable exhibition of "The Painters of Reality." Ignored for nearly three hundred years, one of the greatest of French painters had come back to claim his rights.

Georges de Latour was one of those extroverted visionaries, whose art faithfully reflects certain aspects of the outer world, but reflects them in a state of transfigurement, so that every meanest particular becomes intrinsically significant, a manifestation of the absolute. Most of his compositions are of figures seen by the light of a single candle. A single candle, as Caravaggio and the Spaniards had shown, can give rise to the most enormous theatrical effects. But Latour took no interest in theatrical effects. There is nothing dramatic in his pictures, nothing tragic or pathetic or grotesque, no representation of action, no appeal to the sort of emotions, which people go to the theater to have excited and then appeased. His personages are essentially static. They never *do* anything; they are simply *there* in the same way in which a granite Pharaoh is there, or a bodhisattva from Khmer, or one of Piero's flat-footed angels. And the single candle is used, in every case, to stress this intense but unexcited, impersonal thereness. By exhibiting common things in an uncommon light, its flame makes manifest the living mystery and inexplicable marvel of mere existence. There is so little religiosity in the paintings that in many cases it is impossible to decide whether we are confronted by an illustration to the Bible or a study of models by candlelight. Is the "Nativity" at Rennes *the* nativity, or merely *a* nativity? Is the picture of an old man asleep under the eyes of a young girl merely that?

Or is it of St. Peter in prison being visited by the delivering angel? There is no way of telling. But though Latour's art is wholly without religiosity, it remains profoundly religious in the sense that it reveals, with unexampled intensity, the divine omnipresence.

It must be added that, as a man, this great painter of God's immanence seems to have been proud, hard, intolerably overbearing and avaricious. Which goes to show, yet once more, that there is never a one-to-one correspondence between an artist's work and his character.

V.

At the near point Vuillard painted interiors for the most part, but sometimes also gardens. In a few compositions he managed to combine the magic of propinquity with the magic of remoteness by representing a corner of a room in which there stands or hangs one of his own, or someone else's, representations of a distant view of trees, hills and sky. It is an invitation to make the best of both worlds, the telescopic and the microscopic, at a single glance.

For the rest, I can think of only a very few close-up landscapes by modern European artists. There is a strange "Thicket" by Van Gogh at the Metropolitan. There is Constable's wonderful "Dell in Helmington Park" at the Tate. There is a bad picture, Millais's "Ophelia," made magical, in spite of everything, by its intricacies of summer greenery seen from the point of view, very nearly, of a water rat. And I remember a Delacroix, glimpsed long ago at some loan exhibition, of bark and leaves and blossom at the closest range. There must, of course, be others;

but either I have forgotten, or have never seen them. In any case there is nothing in the West comparable to the Chinese and Japanese renderings of nature at the near point. A spray of blossoming plum, eighteen inches of a bamboo stem with its leaves, tits or finches seen at hardly more than arm's length among the bushes, all kinds of flowers and foliage, of birds and fish and small mammals. Each tiny life is represented as the center of its own universe, the purpose, in its own estimation, for which this world and all that is in it were created; each issues its own specific and individual declaration of independence from human imperialism; each, by ironic implication, derides our absurd pretensions to lay down merely human rules for the conduct of the cosmic game; each mutely repeats the divine tautology: I am that I am.

Nature at the middle distance is familiar—so familiar that we are deluded into believing that we really know what it is all about. Seen very close at hand, or at a great distance, or from an odd angle, it seems disquietingly strange, wonderful beyond all comprehension. The close-up landscapes of China and Japan are so many illustrations of the theme that samsara and nirvana are one, that the Absolute is manifest in every appearance. These great metaphysical, and yet pragmatic, truths were rendered by the Zen-inspired artists of the Far East in yet another way. All the objects of their near-point scrutiny were represented in a state of unrelatedness against a blank of virgin silk or paper. Thus isolated, these transient appearances take on a kind of absolute Thing-in-Itselfhood. Western artists have used this device when painting sacred figures, portraits and, sometimes, natural objects at a distance. Rembrandt's "Mill" and Van Gogh's "Cypresses" are examples of long-range landscapes in which a single feature has been absolutized by isolation. The magical power of many of Goya's etchings, drawings and paintings can be accounted for by the fact that his compositions almost always take the form of a few silhouettes, or even a single silhouette, seen against a blank. These silhouetted shapes possess the visionary quality of intrinsic significance, heightened by isolation and unrelatedness to preternatural intensity. In nature, as in a work of art, the isolation of an object tends to invest it with absoluteness, to endow it with that more-than-symbolic meaning which is identical with being.

> —But there's a Tree—of many, *one*,
> A *single* Field which I have looked upon,
> Both of them speak of something that is gone.

The something which Wordsworth could no longer see

was the "visionary gleam." That gleam, I remember, and that intrinsic significance were the properties of a solitary oak that could be seen from the train, between Reading and Oxford, growing from the summit of a little knoll in a wide expanse of plowland, and silhouetted against the pale northern sky.

The effects of isolation combined with proximity may be studied, in all their magical strangeness, in an extraordinary painting by a seventeenth-century Japanese artist, who was also a famous swordsman and a student of Zen. It represents a butcherbird, perched on the very tip of a naked branch, "waiting without purpose, but in the state of highest tension." Beneath, above and all around is nothing. The bird emerges from the Void, from that eternal namelessness and formlessness, which is yet the very substance of the manifold, concrete and transient universe. That shrike on its bare branch is first cousin to Hardy's wintry thrush. But whereas the thrush insists on teaching us some kind of a lesson, the Far Eastern butcherbird is content simply to exist, to be intensely and absolutely there.

VI.

Many schizophrenics pass most of their time neither on earth, nor in heaven, nor even in hell, but in a gray, shadowy world of phantoms and unrealities. What is true of these psychotics is true, to a lesser extent, of certain neurotics afflicted by a milder form of mental illness. Recently it has been found possible to induce this state of ghostly existence by administering a small quantity of one of the derivatives of adrenalin. For the living, the doors of heaven, hell and limbo are opened, not by "massy keys of metals twain," but by the presence in the blood of one set of chemical compounds and the absence of another set. The shadow world inhabited by some schizophrenics and neurotics closely resembles the world of the dead, as described in some of the earlier religious traditions. Like the wraiths in Sheol and in Homer's Hades, these mentally disturbed persons have lost touch with matter, language and their fellow beings. They have no purchase on life and are condemned to ineffectiveness, solitude and a silence broken only by the senseless squeak and gibber of ghosts.

The history of eschatological ideas marks a genuine progress—a progress which can be described in theological terms as the passage from Hades to Heaven, in chemical terms as the substitution of mescalin and lysergic acid for adrenolutin, and in psychological terms as the advance from catatonia and feelings of unreality to a sense of heightened reality in vision and, finally, in mystical experience.

VII.

Géricault was a negative visionary; for though his art was almost obsessively true to nature, it was true to a nature that had been magically transfigured, in his perceiving and rendering of it, for the worse. "I start to paint a woman," he once said, "but it always ends up as a lion." More often, indeed, it ended up as something a good deal less amiable than a lion—as a corpse, for example, as a demon. His masterpiece, the prodigious "Raft of the *Medusa*," was painted not from life, but from dissolution and decay—from bits of cadavers supplied by medical students, from the emaciated torso and jaundiced face of a friend who was suffering from a disease of the liver. Even the waves on which the raft is floating, even the overarching sky are corpse-colored. It is as though the entire universe had become a dissecting room.

And then there are his demonic pictures. "The Derby," it is obvious, is being run in hell, against a background fairly blazing with darkness visible. "The Horse Startled by Lightning," in the National Gallery, is the revelation, in a single frozen instant, of the strangeness, the sinister and even infernal otherness that hides in familiar things. In the Metropolitan Museum there is a portrait of a child. And what a child! In his luridly brilliant jacket the little darling is what Baudelaire liked to call "a budding Satan," *un Satan en herbe*. And the study of a naked man, also in the Metropolitan, is none other than the budding Satan grown up.

From the acounts which his friends have left of him it is evident that Géricault habitually saw the world about him as a succession of visionary apocalypses. The prancing horse of his early "Officier de Chasseurs" was seen one morning, on the road to Saint-Cloud, in a dusty glare of summer sunshine, rearing and plunging between the shafts of an omnibus. The personages in the "Raft of the *Medusa*" were painted in finished detail, one by one, on the virgin canvas. There was no outline drawing of the whole composition, no gradual building up of an overall harmony of tones and hues. Each particular revelation —of a body in decay, of a sick man in the ghastly extremity of hepatitis—was fully rendered as it was seen and artistically realized. By a miracle of genius, every successive apocalypse was made to fit, prophetically, into a harmonious whole, which existed, when the earlier of the appalling visions were transferred to canvas, only in the artist's imagination.

VIII.

In *Sartor Resartus* Carlyle has left what (in *Mr. Carlyle, My Patient*) his psychosomatic biographer, Dr. James Halliday, calls "an amazing description of a psychotic state of mind, largely depressive, but partly schizophrenic."

"The men and women around me," writes Carlyle, "even speaking too with me, were but Figures; I had practically forgotten that they were alive, that they were not merely automata. Friendship was but an incredible tradition. In the midst of their crowded streets and assemblages I walked solitary; and (except that it was my own heart, not another's, that I kept devouring) savage also as the tiger in the jungle. . . . To me the Universe was all void of Life, of Purpose, of Volition, even of Hostility; it was one huge, dead, immeasurable Steam-Engine, rolling on in its dead indifference, to grind me limb from limb . . . Having no hope, neither had I any definite fear, were it of Man or of Devil. And yet, strangely enough, I lived in a continual, indefinite, pining fear, tremulous, pusillanimous, apprehensive of I knew not what; it seemed as if all things in the Heavens above, and the Earth beneath, would hurt me; as if the Heavens and the Earth were but boundless jaws of a devouring Monster, wherein I, palpitating, waited to be devoured." Renée and the idolater of heroes are evidently describing the same experience. Infinity is apprehended by both, but in the form of "the System," the "immeasurable Steam-Engine." To both, again, all is significant, but negatively significant, so that every event is utterly pointless, every object intensely unreal, every self-styled human being a clockwork dummy, grotesquely going through the motions of work and play, of loving, hating, thinking, of being eloquent, heroic, saintly, what you will—the robots are nothing if not versatile.

About the author

About the book

Read on

Insights,
Interviews
& More...

Aldous Huxley
A Life of the Mind

POET, PLAYWRIGHT, NOVELIST, short
story writer, travel writer, essayist, critic,
philosopher, mystic, and social prophet,
Aldous Huxley was one of the most
accomplished and influential English
literary figures of the mid-twentieth
century. In the course of an extraordinary
prolific writing career, which began in the
early 1920s and continued until his death
in 1963, Huxley underwent a remarkable
process of self-transformation from a derisive
satirist of England's chattering classes to a
deeply religious writer preoccupied with the
human capacity for spiritual transcendence.
Yet in everything Huxley wrote, from the
most frivolous to the most profound, there
runs the common thread of his search to
explain the meaning and possibilities of
human life and perception.

Courtesy of Man Ray/The Granger Collection

Aldous Huxley was born in
Surrey, England, in 1894, the son
of Leonard Huxley, editor of the
prestigious *Cornhill* magazine; and
of Julia Arnold, niece of the poet
and essayist Matthew Arnold, and
sister of Mrs. Humphrey Ward. He
was the grandson of T. H. Huxley,
the scientist. Thus by "birth and
disposition," as one biographer put
it, Huxley belonged to England's
intellectual aristocracy.

As Sybille Bedford writes in her
fascinating biography, *Aldous Huxley*
(Alfred A. Knopf / Harper & Row,
1974): "What we know about him as
a young child is the usual residue of
anecdote and snapshot. During his first years
his head was proportionately enormous, so
that he could not walk till he was two because

he was apt to topple over. 'We put father's hat on him and it fitted.' In another country, at a great distance in time and place, when he lay ill and near his end in southern California, a friend, wanting to distract him, said, 'Aldous, didn't you ever have a nickname when you were small?' and Aldous, who hardly ever talked about his childhood or indeed about himself (possibly because one did not ask) said promptly, 'They called me Ogie. Short for Ogre.'

"The Ogre was a pretty little boy, the photographs . . . show the high forehead, the (then) clear gaze, the tremulous mouth and a sweetness of expression, an alertness beyond that of other angelic little boys looking into a camera. Aldous, his brother, Julian, tells us, sat quietly a good deal of the time 'contemplating the strangeness of things.'

" 'I used to watch him with a pencil,' said his cousin and contemporary Gervas Huxley, 'you see, he was always drawing. . . . My earliest memory of him is sitting—absorbed—to me it was magic, a little boy of my own age drawing so beautifully.'

"He was delicate; he had mischievous moods; he could play. He carried his rag doll about him for company until he was eight. He was fond of grumbling. They gave him a milk mug which bore the inscription: *Oh, isn't the world extremely flat / With nothing whatever to grumble at.*

". . . And Aldous aged six being taken with all the Huxleys to the unveiling of the statue of his grandfather at the Natural History Museum by the Prince of Wales, and his mother trying, in urgent whispers, to persuade Julian, then a young Etonian, to give up his top hat—a very young Etonian and a very new top hat—to Aldous, queasy, overcome, to be sick in."

When Huxley was a sixteen-year-old ▶

> 'They called me Ogie. Short for Ogre.'

student at Eton, he contracted a disease that left him almost totally blind for two years and seriously impaired his vision for years to come. The loss of sight was an "event," Huxley later wrote, "which prevented me from becoming a complete public school English-gentleman." It also ended his early dreams of becoming a doctor. Yet, in a curious way, though he abandoned science for literature, Huxley's outlook remained essentially scientific. As his brother, the zoologist Julian Huxley, wrote, science and mysticism were overlapping and complementary realms in Aldous Huxley's mind: "The more [science] discovers and the more comprehension it gives us of the mechanisms of existence, the more clearly does the mystery of existence itself stand out."

Huxley took his undergraduate degree in literature at Balliol College, Oxford, in 1916, and spent several years during World War I working in a government office. After teaching briefly at Eton, he launched his career as a professional writer in 1920 by taking a job as a drama critic for the *Westminster Gazette*, and a staff writer for *House and Garden* and *Vogue*. Possessed of seemingly infinite literary energy, he wrote poetry, essays, and fiction in his spare time, publishing his first novel, *Crome Yellow*, in 1921. This bright, sharp, mildly shocking satire of upper-class artists won Huxley an immediate reputation as a dangerous wit. He swiftly composed several more novels in a similar vein, including *Antic Hay* (1923) and *Those Barren Leaves* (1925).

In *Point Counter Point* (1928), considered by many critics his strongest novel, Huxley broke new ground, both stylistically and thematically. In a narrative that jumps abruptly from scene to scene and character to character, Huxley confronts modern man's disillusionment with religion, art, sex, and politics. The character Philip Quarles, a novelist intent on "transform[ing] a detached intellectual skepticism into a way of harmonious all-round living," is the closest Huxley came to painting his own portrait in fiction. *Brave New World* (1932), though less experimental in style than *Point Counter Point*, is more radical in its pessimistic view of human nature. Huxley's antiutopia, with its eerie combination of totalitarian government and ubiquitous feel-good drugs and sex, disturbed many readers of his day; but it has proven to be his most enduring and influential work.

During the 1930s, Huxley turned increasingly toward an exploration of fundamental questions of philosophy, sociology, politics, and ethics. In his 1936 novel *Eyeless in Gaza* he wrote of a man's transformation from cynic to mystic, and as war threatened Europe once again, he allied himself with the pacifist movement and began lecturing widely on peace and internationalism.

For a number of years Huxley lived in Italy, where he formed a close relationship with D. H. Lawrence, whose letters he edited in 1933. In 1937, Huxley and his Belgian-born wife, Maria Nys, and their son, Matthew, left Europe to live in Southern California for the rest of his life. Maria Huxley died of cancer in 1955, and the following year Huxley married the Italian violinist and psychotherapist Laura Archera.

In the 1940s and 1950s, Huxley changed direction yet again as he became fascinated by the spiritual life, in particular with the possibility of direct communication between people and the divinity. Huxley read widely in the writings of the mystics and assembled an anthology of mystical writing called *The Perennial Philosophy* (1945). Around this time he began experimenting with mind-altering drugs like mescaline and LSD, which he came to believe gave users essentially the same experiences that mystics attained through fasting, prayer, and meditation. *The Doors of Perception* (1954) and *Heaven and Hell* (1956), Huxley's books about the effects of what he termed psychedelic drugs, became essential texts for the counterculture during the 1960s. Yet Huxley's brother Julian cautions against the image of Aldous as a kind of spiritual godfather to hippies: "One of Aldous's major preoccupations was how to achieve self-transcendence while yet remaining a committed social being—how to escape from the prison bars of self and the pressures of here and now into realms of pure goodness and pure enjoyment."

Huxley pursued his quest for "pure goodness and pure enjoyment" right up to the end of his life on November 22, 1963. Today he is remembered as one of the great explorers of twentieth-century literature, a writer who continually reinvented himself as he pushed his way deeper and deeper into the mysteries of human consciousness.

Drugs That Shape Men's Minds

An Essay by Aldous Huxley

IN THE COURSE OF HISTORY many more people have died for their drink and their dope than have died for their religion or their country. The craving for ethyl alcohol and the opiates has been stronger, in these millions, than the love of God, of home, of children; even of life. Their cry was not for liberty or death; it was for death preceded by enslavement. There is a paradox here, and a mystery. Why should such multitudes of men and women be so ready to sacrifice themselves for a cause so utterly hopeless and in ways so painful and so profoundly humiliating?

To this riddle there is, of course, no simple or single answer. Human beings are immensely complicated creatures, living simultaneously in a half dozen different worlds. Each individual is unique and, in a number of respects, unlike all the other members of the species. None of our motives is unmixed, none of our actions can be traced back to a single source and, in any group we care to study, behavior patterns that are observably similar may be the result of many constellations of dissimilar causes.

Thus, there are some alcoholics who seem to have been biochemically predestined to alcoholism. (Among rats, as Prof. Roger Williams, of the University of Texas, has shown, some are born drunkards; some are born teetotalers and will never touch the stuff.) Other alcoholics have been foredoomed not by some inherited defect in their biochemical make-up, but by their neurotic reactions to distressing events in their childhood or adolescence. Again, others embark upon their course of slow suicide as a result of mere imitation and good fellowship because they have made such an "excellent adjustment to their group"—a process which, if the group happens to be criminal, idiotic or merely ignorant, can bring only disaster to the well-adjusted individual. Nor must we forget that large class of addicts who have taken to drugs or drink in order to escape from physical pain. Aspirin, let us remember, is a very recent invention. Until late in the Victorian era, "poppy and mandragora," along with henbane and ethyl alcohol, were the only pain relievers available to civilized man. Toothache, arthritis and neuralgia could, and frequently did, drive men and women to become opium addicts.

De Quincey, for example, first resorted to opium in order to relieve "excruciating rheumatic pains of the head." He swallowed his poppy and, an hour later, "What a resurrection from the lowest depths of the inner spirit! What an apocalypse!" And it was not merely that he felt no more pain. "This negative effect was swallowed up in the immensity of those positive effects which had opened up before me, in the abyss of divine enjoyment thus suddenly revealed. . . . Here was the secret of happiness, about which the philosophers had disputed for so many ages, at once discovered."

"Resurrection, apocalypse, divine enjoyment, happiness. . . ." De Quincey's words lead us to the very heart of our paradoxical mystery. The problem of drug addiction and excessive drinking is not merely a matter of chemistry and psychopathology, of relief from pain and conformity with a bad society. It is also a problem in metaphysics—a problem, one might almost say, in theology. In ▶

66 Toothache, arthritis and neuralgia could, and frequently did, drive men and women to become opium addicts. 99

The Varieties of Religious Experience, William James has touched on these metaphysical aspects of addiction:

> The sway of alcohol over mankind is unquestionably due to
> its power to stimulate the mystical faculties in human nature,
> usually crushed to earth by the cold facts and dry criticisms of
> the sober hour. Sobriety diminishes, discriminates and says no.
> Drunkenness expands, unites and says yes. It is in fact the great
> exciter of the Yes function in man. It brings its votary from the
> chill periphery of things into the radiant core. It makes him for
> the moment one with truth. Not through mere perversity do
> men run after it. To the poor and the unlettered it stands in the
> place of symphony concerts and literature; and it is part of the
> deeper mystery and tragedy of life that whiffs and gleams of
> something that we immediately recognize as excellent should
> be vouchsafed to so many of us only through the fleeting earlier
> phases of what, in its totality, is so degrading a poison. The
> drunken consciousness is one bit of the mystic consciousness,
> and our total opinion of it must find its place in our opinion
> of that larger whole.

William James was not the first to detect a likeness between drunkenness and the mystical and pre-mystical states. On the day of Pentecost there were people who explained the strange behavior of the disciples by saying, "These men are full of new wine."

Peter soon undeceived them: "These are not drunken, as ye suppose, seeing it is but the third hour of the day. But this is that which was spoken by the prophet Joel. And it shall come to pass in the last days, saith God, I will pour out of my Spirit upon all flesh."

And it is not only by "the dry critics of the sober hour" that the state of God-intoxication has been likened to drunkenness. In their efforts to express the inexpressible, the great mystics themselves have done the same. Thus, St. Theresa of Avila tells us that she "regards the center of our soul as a cellar, into which God admits us as and when it pleases Him, so as to intoxicate us with the delicious wine of His grace."

Every fully developed religion exists simultaneously on several different levels. It exists as a set of abstract concepts about the world and its governance. It exists as a set of rites and sacraments, as a traditional method for manipulating the symbols, by means of which beliefs about the cosmic order are expressed. It exists as the feelings of love, fear and devotion evoked by this manipulation of symbols.

And finally it exists as a special kind of feeling or intuition—a sense of the oneness of all things in their divine principle, a realization (to use the language of Hindu theology) that "thou art That," a mystical experience of what seems self-evidently to be union with God.

The ordinary waking consciousness is a very useful and, on most occasions, an indispensable state of mind; but it is by no means the only form of consciousness, nor in all circumstances the best. Insofar as he transcends his ordinary self and his ordinary mode of awareness, the mystic is able to enlarge his vision, to look more deeply into the unfathomable miracle of existence.

The mystical experience is doubly valuable; it is valuable because it gives the experiencer a better understanding of himself and the world and because it may help him to lead a less self-centered and more creative life.

In hell, a great religious poet has written, the punishment of the lost is to be "their sweating selves, but worse." On earth we are not worse than we are; we are merely our sweating selves, period.

Alas, that is quite bad enough. We love ourselves to the point of idolatry; but we also intensely dislike ourselves—we find ourselves unutterably boring. Correlated with this distaste for the idolatrously worshiped self, there is in all of us a desire, sometimes latent, sometimes conscious and passionately expressed, to escape from the prison of our individuality, an urge to self-transcendence. It is to this urge that we owe mystical theology, spiritual exercises and yoga—to this, too, that we owe alcoholism and drug addiction.

Modern pharmacology has given us a host of new synthetics, but in the field of the naturally occurring mind changers it has made no radical discoveries. All the botanical sedatives, stimulants, vision revealers, happiness promoters and cosmic-consciousness arousers were found out thousands of years ago, before the dawn of history.

In many societies at many levels of civilization attempts have been made to fuse drug intoxication with God intoxication. In ancient Greece, for example, ethyl alcohol had its place in the established religion. Dionysus, or Bacchus, as he was often called, was a true divinity. His worshipers addressed him as *Lusios*, "Liberator," or as *Theoinos*, "God-wine." The latter name telescopes fermented grape juice and the supernatural into a single pentecostal experience. "Born a god," writes Euripides, "Bacchus is poured out as a libation to the gods, and through him men receive good." Unfortunately they also receive harm. The blissful experience of self-transcendence which alcohol makes possible has to be paid for, and the price is exorbitantly high. ▸

Drugs That Shape Men's Minds (*continued*)

Complete prohibition of all chemical mind changers can be decreed, but cannot be enforced, and tends to create more evils than it cures. Even more unsatisfactory has been the policy of complete toleration and unrestricted availability. In England, during the first years of the eighteenth century, cheap untaxed gin—"drunk for a penny, dead drunk for two-pence"—threatened society with complete demoralization. A century later, opium, in the form of laudanum, was reconciling the victims of the Industrial Revolution to their lot—but at an appalling cost in terms of addiction, illness and early death. Today most civilized societies follow a course between the two extremes of total prohibition and total toleration. Certain mind-changing drugs, such as alcohol, are permitted and made available to the public on payment of a very high tax, which tends to restrict their consumption. Other mind changers are unobtainable except under doctors' orders—or illegally from a dope pusher. In this way the problem is kept within manageable bounds. It is most certainly not solved. In their ceaseless search for self-transcendence, millions of would-be mystics become addicts, commit scores of thousands of crimes and are involved in hundreds of thousands of avoidable accidents.

Do we have to go on in this dismal way indefinitely? Up until a few years ago, the answer to such a question would have been a rueful "Yes, we do." Today, thanks to recent developments in biochemistry and pharmacology, we are offered a workable alternative. We see that it may soon be possible for us to do something better in the way of chemical self-transcendence than what we have been doing so ineptly for the last seventy or eighty centuries.

Is it possible for a powerful drug to be completely harmless? Perhaps not. But the physiological cost can certainly be reduced to the point where it becomes negligible. There are powerful mind changers which do their work without damaging the taker's psychophysical organism and without inciting him to behave like a criminal or a lunatic. Biochemistry and pharmacology are just getting into their stride. Within a few years there will probably be dozens of powerful but—physiologically and socially speaking— very inexpensive mind changers on the market.

In view of what we already have in the way of powerful but nearly harmless drugs; in view, above all, of what unquestionably we are very soon going to have—we ought to start immediately to give some serious thought to the problem of the new mind changers. How ought they to be used? How can they be abused? Will human beings be better and happier for their discovery? Or worse and more miserable?

The matter requires to be examined from many points of view. It is simultaneously a question for biochemists and physicians, for psychologists and social anthropologists, for legislators and law-enforcement officers. And finally it is an ethical question and a religious question. Sooner or later—and the sooner, the better—the various specialists concerned will have to meet, discuss and then decide, in the light of the best available evidence and the most imaginative kind of foresight, what should be done. Meanwhile let us take a preliminary look at this many-faceted problem.

Last year American physicians wrote 48,000,000 prescriptions for tranquilizing drugs, many of which have been refilled, probably more than once. The tranquilizers are the best known of the new, nearly harmless mind changers. They can be used by most people, not indeed with complete impunity, but at a reasonably low physiological cost. Their enormous popularity bears witness to the fact that a great many people dislike both their environment and "their sweating selves." Under tranquilizers the degree of their self-transcendence is not very great; but it is enough to make all the difference, in many cases, between misery and contentment.

In theory, tranquilizers should be given only to persons suffering from rather severe forms of neurosis or psychosis. In practice, unfortunately, many physicians have been carried away by the current pharmacological fashion and are prescribing tranquilizers to all and sundry. The history of medical fashions, it may be remarked, is at least as grotesque as the history of fashions in women's hats—at least as grotesque and, since human lives are at stake, considerably more tragic. In the present case, millions of patients who had no real need of the tranquilizers have been given the pills by their doctors and have learned to resort to them in every predicament, however triflingly uncomfortable. This is very bad medicine and, from the pill taker's point of view, dubious morality and poor sense.

There are circumstances in which even the healthy are justified in resorting to the chemical control of negative emotions. If you really can't keep your temper, let a tranquilizer keep it for you. But for healthy people to resort to a chemical mind changer every time they feel annoyed or anxious or tense is neither sensible nor right. Too much tension and anxiety can reduce a man's efficiency—but so can too little. There are many occasions when it is entirely proper for us to feel concerned, when an excess of placidity might reduce our chances of dealing effectively with a ticklish situation. On these occasions, tension mitigated and directed from within by the psychological methods of self-control is ▶

preferable from every point of view to complacency imposed from
without by the methods of chemical control.

And now let us consider the case—not, alas, a hypothetical case—
of two societies competing with each other. In Society A, tranquilizers
are available by prescription and at a rather stiff price—which means,
in practice, that their use is confined to that rich and influential minority
which provides the society with its leadership. This minority of leading
citizens consumes several billions of the complacency-producing pills
every year. In Society B, on the other hand, the tranquilizers are not so
freely available, and the members of the influential minority do not
resort, on the slightest provocation, to the chemical control of what
may be necessary and productive tension. Which of these two competing
societies is likely to win the race? A society whose leaders make an
excessive use of soothing syrups is in danger of falling behind a
society whose leaders are not overtranquilized.

Now let us consider another kind of drug—still undiscovered, but
probably just around the corner—a drug capable of making people feel
happy in situations where they would normally feel miserable. Such a
drug would be a blessing, but a blessing fraught with grave political
dangers. By making harmless chemical euphoria freely available,
a dictator could reconcile an entire population to a state of affairs
to which self-respecting human beings ought not to be reconciled.
Despots have always found it necessary to supplement force by political
or religious propaganda. In this sense the pen is mightier than the sword.
But mightier than either the pen or the sword is the pill. In mental
hospitals it has been found that chemical restraint is far more effective
than strait jackets or psychiatry. The dictatorships of tomorrow will
deprive men of their freedom, but will give them in exchange a happiness
none the less real, as a subjective experience, for being chemically
induced. The pursuit of happiness is one of the traditional rights
of man; unfortunately, the achievement of happiness may turn out
to be incompatible with another of man's rights—namely, liberty.

It is quite possible, however, that pharmacology will restore with one
hand what it takes away with the other. Chemically induced euphoria
could easily become a threat to individual liberty; but chemically induced
vigor and chemically heightened intelligence could easily be liberty's
strongest bulwark. Most of us function at about 15 per cent of capacity.
How can we step up our lamentably low efficiency?

Two methods are available—the educational and the biochemical.
We can take adults and children as they are and give them a much better
training than we are giving them now. Or, by appropriate biochemical

methods, we can transform them into superior individuals. If these superior individuals are given a superior education, the results will be revolutionary. They will be startling even if we continue to subject them to the rather poor educational methods at present in vogue.

Will it in fact be possible to produce superior individuals by biochemical means? The Russians certainly believe it. They are now halfway through a Five Year Plan to produce "pharmacological substances that normalize higher nervous activity and heighten human capacity for work." Precursors of these future mind improvers are already being experimented with. It has been found, for example, that when given in massive doses some of the vitamins—nicotinic acid and ascorbic acid for example—sometimes produce a certain heightening of psychic energy. A combination of two enzymes—ethylene disulphonate and adenosine triphosphate, which, when injected together, improve carbohydrate metabolism in nervous tissue—may also turn out to be effective.

Meanwhile good results are being claimed for various new synthetic, nearly harmless stimulants. There is iproniazid, which, according to some authorities, "appears to increase the total amount of psychic energy." Unfortunately, iproniazid in large doses has side effects which in some cases may be extremely serious! Another psychic energizer is an amino alcohol which is thought to increase the body's production of acetylcholine, a substance of prime importance in the functioning of the nervous system. In view of what has already been achieved, it seems quite possible that, within a few years, we may be able to lift ourselves up by our own biochemical bootstraps.

In the meantime let us all fervently wish the Russians every success in their current pharmacological venture. The discovery of a drug capable of increasing the average individual's psychic energy, and its wide distribution throughout the U.S.S.R., would probably mean the end of Russia's present form of government. Generalized intelligence and mental alertness are the most powerful enemies of dictatorship and at the same time the basic conditions of effective democracy. Even in the democratic West we could do with a bit of psychic energizing. Between them, education and pharmacology may do something to offset the effects of that deterioration of our biological material to which geneticists have frequently called attention.

From these political and ethical considerations let us now pass to the strictly religious problems that will be posed by some of the new mind changers. We can foresee the nature of these future problems by studying the effects of a natural mind changer, which has been used ▶

Drugs That Shape Men's Minds *(continued)*

for centuries past in religious worship; I refer to the peyote cactus of Northern Mexico and the Southwestern United States. Peyote contains mescaline—which can now be produced synthetically—and mescaline, in William James' phrase, "stimulates the mystical faculties in human nature" far more powerfully and in a far more enlightening way than alcohol and, what is more, it does so at a physiological and social cost that is negligibly low. Peyote produces self-transcendence in two ways— it introduces the taker into the Other World of visionary experience, and it gives him a sense of solidarity with his fellow worshipers, with human beings at large and with the divine nature of things.

The effects of peyote can be duplicated by synthetic mescaline and by LSD (lysergic acid diethylamide), a derivative of ergot. Effective in incredibly small doses, LSD is now being used experimentally by psychotherapists in Europe, in South America, in Canada and the United States. It lowers the barrier between conscious and subconscious and permits the patient to look more deeply and understandingly into the recesses of his own mind. The deepening of self-knowledge takes place against a background of visionary and even mystical experience.

When administered in the right kind of psychological environment, these chemical mind changers make possible a genuine religious experience. Thus a person who takes LSD or mescaline may suddenly understand—not only intellectually but organically, experientially— the meaning of such tremendous religious affirmations as "God is love," or "Though He slay me, yet will I trust in Him."

It goes without saying that this kind of temporary self-transcendence is no guarantee of permanent enlightenment or a lasting improvement of conduct. It is a "gratuitous grace," which is neither necessary nor sufficient for salvation, but which if properly used, can be enormously helpful to those who have received it. And this is true of all such experiences, whether occurring spontaneously, or as the result of swallowing the right kind of chemical mind changer, or after undertaking a course of "spiritual exercises" or bodily mortification.

Those who are offended by the idea that the swallowing of a pill may contribute to a genuinely religious experience should remember that all the standard mortifications—fasting, voluntary sleeplessness and self-torture—inflicted upon themselves by the ascetics of every religion for the purpose of acquiring merit, are also, like the mind-changing drugs, powerful devices for altering the chemistry of the body in general and the nervous system in particular. Or consider the procedures generally known as spiritual exercises. The breathing techniques taught by the yogi of India result in prolonged suspensions of respiration. These in turn result in an increased concentration of carbon dioxide in the blood; and the psychological consequence of this is a change in the quality of consciousness. Again, meditations involving long, intense concentration upon a single idea or image may also result—for neurological reasons which I do not profess to understand—in a slowing down of respiration and even in prolonged suspensions of breathing.

Many ascetics and mystics have practiced their chemistry-changing mortifications and spiritual exercises while living, for longer or shorter periods, as hermits. Now, the life of a hermit, such as Saint Anthony, is a life in which there are very few external stimuli. But as Hebb, John Lilly and other experimental psychologists have recently shown in the laboratory, a person in a limited environment, which provides very few external stimuli, soon undergoes a change in the quality of his consciousness and may transcend his normal self to the point of hearing voices or seeing visions, often extremely unpleasant, like so many of Saint Anthony's visions, but sometimes beatific.

That men and women can, by physical and chemical means, transcend themselves in a genuinely spiritual way is something which, to the squeamish idealist, seems rather shocking. But, after all, the drug or the physical exercise is not the cause of the spiritual experience; it is only its occasion.

Writing of William James' experiments with nitrous oxide, Bergson has summed up the whole matter in a few lucid sentences. "The psychic disposition was there, potentially, only waiting a signal to express itself in action. It might have been evoked spiritually by an effort made on its own spiritual level. But it could just as well be brought about materially, by an inhibition of what inhibited it, by the removing of an obstacle; and this effect was the wholly negative one produced by the drug." Where, for any reason, physical or moral, the psychological dispositions are unsatisfactory, the removal of obstacles by a drug or by ascetic practices will result in a negative rather than a positive spiritual experience. Such an infernal experience is extremely distressing, but may also be extremely salutary. There are plenty of people to whom a few hours in hell—the hell that they themselves have done so much to create—could do a world of good.

Physiologically costless, or nearly costless, stimulators of the mystical faculties are now making their appearance, and many kinds of them will soon be on the market. We can be quite sure that, as and when they become available, they will be extensively used. The urge to self-transcendence is so strong and so general that it cannot be otherwise. ▶

Drugs That Shape Men's Minds (*continued*)

In the past, very few people have had spontaneous experiences of a pre-mystical or fully mystical nature; still fewer have been willing to undergo the psychophysical disciplines which prepare an insulated individual for this kind of self-transcendence. The powerful but nearly costless mind changers of the future will change all this completely. Instead of being rare, pre-mystical and mystical experiences will become common. What was once the spiritual privilege of the few will be made available to the many. For the ministers of the world's organized religions, this will raise a number of unprecedented problems. For most people, religion has always been a matter of traditional symbols and of their own emotional, intellectual and ethical response to those symbols. To men and women who have had direct experience of self-transcendence into the mind's Other World of vision and union with the nature of things, a religion of mere symbols is not likely to be very satisfying. The perusal of a page from even the most beautifully written cookbook is no substitute for the eating of dinner. We are exhorted to "*taste* and see that the Lord is good."

In one way or another, the world's ecclesiastical authorities will have to come to terms with the new mind changers. They may come to terms with them negatively, by refusing to have anything to do with them. In that case, a psychological phenomenon, potentially of great spiritual value, will manifest itself outside the pale of organized religion. On the other hand, they may choose to come to terms with the mind changers in some positive way—exactly how, I am not prepared to guess.

My own belief is that, though they may start by being something of an embarrassment, these new mind changers will tend in the long run to deepen the spiritual life of the communities in which they are available. That famous "revival of religion," about which so many people have been talking for so long, will not come about as the result of evangelistic mass meetings or the television appearances of photogenic clergymen. It will come about as the result of biochemical discoveries that will make it possible for large numbers of men and women to achieve a radical self-transcendence and a deeper understanding of the nature of things. And this revival of religion will be at the same time a revolution. From being an activity mainly concerned with symbols, religion will be transformed into an activity concerned mainly with experience and intuition—an everyday mysticism underlying and giving significance to everyday rationality, everyday tasks and duties, everyday human relationships. 〰

Selected from Collected Essays *by Aldous Huxley, Harper & Brothers, 1958.*

The Complete Aldous Huxley Bibliography

Dates are the year of first publication.

The Burning Wheel	1916
Jonah	1917
The Defeat of Youth and Other Poems	1918
Leda	1920
Limbo: Notes and Essays	1920
Crome Yellow	1921
Mortal Coils: Five Stories	1922
On the Margin	1923
Antic Hay	1923
Little Mexican	1924
Those Barren Leaves	1925
Along the Road: Notes and Essays	1925
Two or Three Graces: Four Stories	1926
Jesting Pilate: An Intellectual Holiday (The Diary of a Journey)	1926
Essays New and Old (U.S. title: Essays Old and New)	1926
Proper Studies	1927
Point Counter Point	1928
Do What You Will: Essays	1929
Brief Candles	1930
Vulgarity in Literature and Other Essays: Digressions from a Theme	1930
The World of Light	1931
The Cicadas and Other Poems	1931
Music at Night and Other Essays	1931
Brave New World	1932
Texts and Pretexts: An Anthology of Commentaries	1932
Beyond the Mexique Bay	1934
Eyeless in Gaza	1936

Read on

The Complete Aldous Huxley Bibliography
(continued)

Have You Read?
More by Aldous Huxley

THE DEVILS OF LOUDUN

First published in 1952, *The Devils of Loudun* is Aldous Huxley's thrilling account of one of history's most sensational cases of mass demonic possession. The year 1643: When an entire convent is apparently possessed by the devil, a charismatic priest is accused of being in league with Satan and seducing the nuns—both spiritually and sexually. After a celebrated trial, the priest, Urban Grandier, was burnt at the stake for witchcraft. Here is the gripping true history of Grandier and the nuns of Loudun, as told by one of the master storytellers of the twentieth century.

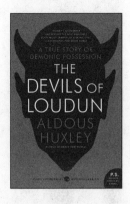

"Huxley's masterpiece and perhaps the most enjoyable book about spirituality ever written. In telling the grotesque, bawdy and true story of a seventeenth-century convent of cloistered French nuns who contrived to have a priest they never met burned alive as a warlock . . . Huxley painlessly conveys a wealth of information about mysticism and the unconscious."
                                      —*Washington Post Book World*

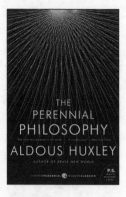

THE PERENNIAL PHILOSOPHY

An astonishing collection of writings drawn from the world's great religions, edited and commented upon by Huxley with characteristic insight, wit, and passion.

"It is the masterpiece of all anthologies. As Mr. Huxley has proved before, he can find and frame rare beauty in literature, and here, long before Freud, writers are quoted who combine beauty with proud psychology."
—*New York Times*

BRAVE NEW WORLD

The astonishing novel *Brave New World*, originally published in 1932, presents Aldous Huxley's vision of the future— of a world utterly transformed. Through the most efficient scientific and psychological engineering, people are genetically designed to be passive and therefore consistently useful to the ruling class. This powerful work of speculative fiction sheds a blazing critical light on the present and is considered to be Aldous Huxley's most enduring masterpiece.

"Mr. Huxley is eloquent in his declaration of an artist's faith in man, and it is his eloquence, bitter in attack, noble in defense, that, when one has closed the book, one remembers." —*Saturday Review of Literature*

"Huxley never went out of style. Something about his work seem[s] to tug at our consciousness. . . . There is no escape from anxiety and struggle, and Huxley assists us in attaining this valuable glimpse of the obvious, precisely because it was a conclusion that was in many ways unwelcome to him."
—Christopher Hitchens